Praise for *The Soul*

"*The Soul of an Octopus* does
for Hawk did for raptors."

"Luminous. . . . An animal a
on every page . . . spellbindir

"Joyful."

—*Library Journal* Editors' Spring Pick

"Delightful."

—*Nature*

"Compassionate, wise, and tender . . ."

—*St. Paul Pioneer Press*

"Touching . . . informative . . . entertaining."

—*The Daily Beast*

"[Montgomery's] compassion and respect for the species make for a buoy-
ing read."

—*Newsday*

"Exceptionally affecting and enlightening . . . inspired . . . rigorous scientific curiosity and enraptured wonder and empathy for all living beings . . .
gripping and entwining . . . funny and moving . . . uniquely intimate."

—*Booklist* (starred review)

"Treat[s] each octopus like a character in a Jane Austen novel."

—*Vox*

"An astoundingly beautiful read . . . scientifically illuminating and deeply
poetic. . . . A worthy addition to the best science books of the year."

—*Science Friday*, National Public Radio

"Journalistic immersion . . . add[s] to our understanding of ourselves."

—*New Scientist*

"Reveals a watery world of animal intelligence painted with such sparkle that I was bereft to leave it for dry land."

—*Times Literary Supplement*

"Montgomery's passion . . . is infectious . . . her awe and admiration are uplifting. . . ."

—*Union Leader*

"Miraculously insightful and enchanting . . . fascinating . . . riveting and ablaze with rigor . . . bewitching prose. . . . A literary naturalist who paints the marvels of the ocean's depths like Thoreau did the marvels of the New England woods. Astoundingly beautiful . . . scientifically illuminating and deeply poetic. . . . A worthy addition to the best science books of the year."

—*Brainpickings*

"A fascinating glimpse into an alien consciousness."

—*Kirkus Reviews*

"Gripping."

—*Global Newswire*

"Charming and moving . . . with extraordinary scientific research."

—*The Guardian* (UK)

"Encourages the reader to reflect on his or her own definition of consciousness and 'soul'."

—*Science*

"Diving deeper than Jules Verne ever dreamed, *The Soul of an Octopus* is a page-turning adventure that will leave you breathless."

—Vicki Constantine Croke, author of *Elephant Company*

"Montgomery immerses readers into an intriguing, seductive world . . . beautifully written."

—Vint Virga, DVM, author of *The Soul of All Living Creatures*

"A captivating book on an intelligence as 'alien' as one from outer space. And it's not science fiction."

—Bernd Heinrich, author of *Mind of the Raven*

the
Soul of an Octopus

A Surprising
Exploration into
the Wonder of Consciousness

Sy Montgomery

ATRIA PAPERBACK

NEW YORK · LONDON · TORONTO · SYDNEY · NEW DELHI

ATRIA PAPERBACK

An Imprint of Simon & Schuster, Inc.
1230 Avenue of the Americas
New York, NY 10020

First Atria Paperback edition April 2016

ATRIA PAPERBACK and colophon are trademarks of Simon & Schuster, Inc.

For information about special discounts for bulk purchases, please contact
Simon & Schuster Special Sales at 1-866-506-1949
or business@simonandschuster.com.

The Simon & Schuster Speakers Bureau can bring authors to your
live event. For more information or to book an event contact the
Simon & Schuster Speakers Bureau at 1-866-248-3049
or visit our website at www.simonspeakers.com.

Interior design by Paul Dippolito

Manufactured in the United States of America

30   29   28   27   26   25

The Library of Congress has cataloged the hardcover edition as follows:

Montgomery, Sy.
    The soul of an octopus : a surprising exploration into the wonder of
consciousness / Sy Montgomery.
        pages cm
Includes bibliographical references and index.
1. Octopuses—Behavior. I. Title.
QL430.3O2M66 2015
594'.56—dc23
    2014038751

ISBN 978-1-4516-9771-1
ISBN 978-1-4516-9772-8 (pbk)
ISBN 978-1-4516-9774-2 (ebook)

For Anna
"Yesterday remains perfect"

Contents

CHAPTER ONE

Athena

Encountering the Mind of a Mollusk

On a rare, warm day in mid-March, when the snow was melting into mud in New Hampshire, I traveled to Boston, where everyone was strolling along the harbor or sitting on benches licking ice cream cones. But I quit the blessed sunlight for the moist, dim sanctuary of the New England Aquarium. I had a date with a giant Pacific octopus.

I knew little about octopuses—not even that the scientifically correct plural is not *octopi*, as I had always believed (it turns out you can't put a Latin ending—*i*—on a word derived from Greek, such as *octopus*). But what I did know intrigued me. Here is an animal with venom like a snake, a beak like a parrot, and ink like an old-fashioned pen. It can weigh as much as a man and stretch as long as a car, yet it can pour its baggy, boneless body through an opening the size of an orange. It can change color and shape. It can taste with its skin. Most fascinating of all, I had read that octopuses are smart. This bore out what scant experience I had already had; like many who visit octopuses in public aquariums, I've often had the feeling that the octopus I was watching was watching me back, with an interest as keen as my own.

How could that be? It's hard to find an animal more unlike a human than an octopus. Their bodies aren't organized like ours. We go: head, body, limbs. They go: body, head, limbs. Their mouths are in their armpits—or, if you prefer to liken their arms to our lower, instead of upper, extremities, between their legs. They breathe water. Their appendages are covered with dexterous, grasping suckers, a structure for which no mammal has an equivalent.

And not only are octopuses on the opposite side of the great vertebral divide that separates the backboned creatures such as mammals, birds, reptiles, amphibians, and fish from everything else; they are classed within the invertebrates as mollusks, as are slugs and snails and clams, animals that are not particularly renowned for their intellect. Clams don't even have brains.

More than half a billion years ago, the lineage that would lead to octopuses and the one leading to humans separated. Was it possible, I wondered, to reach another mind on the other side of that divide?

Octopuses represent the great mystery of the Other. They seem completely alien, and yet their world—the ocean—comprises far more of the Earth (70 percent of its surface area; more than 90 percent of its habitable space) than does land. Most animals on this planet live in the ocean. And most of them are invertebrates.

I wanted to meet the octopus. I wanted to touch an alternate reality. I wanted to explore a different kind of consciousness, if such a thing exists. What is it like to be an octopus? Is it anything like being a human? Is it even possible to know?

So when the aquarium's director of public relations met me in the lobby and offered to introduce me to Athena, the octopus, I felt like a privileged visitor to another world. But what I began to discover that day was my own sweet blue planet—a world breathtakingly alien, startling, and wondrous; a place where, after half a century of life on this earth, much of it as a naturalist, I would at last feel fully at home.

❊

Athena's lead keeper isn't in. My heart sinks; not just anyone can open up the octopus tank, and for good reason. A giant Pacific octopus—the largest of the world's 250 or so octopus species—can easily overpower a person. Just one of a big male's three-inch-diameter suckers can lift 30 pounds, and a giant Pacific octopus has 1,600 of them. An octopus bite can inject a neurotoxic venom as well as saliva that has the ability to dissolve flesh. Worst of all, an octopus can take the opportunity to escape from an open tank, and an escaped octopus is a big problem for both the octopus and the aquarium.

Happily, another aquarist, Scott Dowd, will help me. A big guy in his early forties with a silvery beard and twinkling blue eyes, Scott is the senior aquarist for the Freshwater Gallery, which is down the hall from Cold Marine, where Athena lives. Scott first came to the aquarium as a baby in diapers on its opening day, June 20, 1969, and basically never left. He knows almost every animal in the aquarium personally.

Athena is about two and a half years old and weighs roughly 40 pounds, Scott explains, as he lifts the heavy lid covering her tank. I mount the three short steps of a small movable stair and lean over to see. She stretches about five feet long. Her head—by "head," I mean both the actual head and the mantle, or body, because that's where we mammals expect an animal's head to be—is about the size of a small watermelon. "Or at least a honeydew," says Scott. "When she first came, it was the size of a grapefruit." The giant Pacific octopus is one of the fastest-growing animals on the planet. Hatching from an egg the size of a grain of rice, one can grow both longer and heavier than a man in three years.

By the time Scott has propped open the tank cover, Athena has already oozed from the far corner of her 560-gallon tank to investigate us. Holding to the corner with two arms, she unfurls the others, her whole body red with excitement, and reaches to the surface. Her white suckers face up, like the palm of a person reaching out for a handshake.

"May I touch her?" I ask Scott.

"Sure," he says. I take off my wristwatch, remove my scarf, roll up my sleeves, and plunge both arms elbow-deep into the shockingly cold 47°F water.

Twisting, gelatinous, her arms boil up from the water, reaching for mine. Instantly both my hands and forearms are engulfed by dozens of soft, questing suckers.

Not everyone would like this. The naturalist and explorer William Beebe found the touch of the octopus repulsive. "I have always a struggle before I can make my hands do their duty and seize a tentacle," he confessed. Victor Hugo imagined such an event as an unmitigated horror leading to certain doom. "The spectre lies upon you; the tiger can only devour you; the devil-fish, horrible, sucks your life-blood away," Hugo wrote in *Toilers of the Sea*. "The muscles swell, the fibres of the body are contorted, the skin cracks under the loathsome oppression, the blood spurts out and mingles horribly with the lymph of the monster, which clings to the victim with innumerable hideous mouths. . . ." Fear of the octopus lies deep in the human psyche. "No animal is more savage in causing the death of man in the water," Pliny the Elder wrote in *Naturalis Historia*, circa AD 79, "for it struggles with him by coiling round him and it swallows him with sucker-cups and drags him asunder. . . ."

But Athena's suction is gentle, though insistent. It pulls me like an alien's kiss. Her melon-size head bobs to the surface, and her left eye—octopuses have a dominant eye, as people have dominant hands—swivels in its socket to meet mine. Her black pupil is a fat hyphen in a pearly globe. Its expression reminds me of the look in the eyes of paintings of Hindu gods and goddesses: serene, all-knowing, heavy with wisdom stretching back beyond time.

"She's looking right at you," Scott says.

As I hold her glittering gaze, I instinctively reach to touch her head. "As supple as leather, as tough as steel, as cold as night," Hugo wrote of the octopus's flesh; but to my surprise, her head is silky and softer than custard. Her skin is flecked with ruby and silver, a night sky reflected on the wine-dark sea. As I stroke her with my fingertips, her skin goes white beneath my touch. White is the color of a relaxed octopus; in cuttlefish, close relatives of octopus, females turn white when they encounter a fellow female, someone whom they need not fight or flee.

It is possible that Athena, in fact, knows I am female. Female octopuses, like female humans, possess estrogen; she could be tasting and recognizing mine. Octopuses can taste with their entire bodies, but this sense is most exquisitely developed in their suckers. Athena's is an exceptionally intimate embrace. She is at once touching and tasting my skin, and possibly the muscle, bone, and blood beneath. Though we have only just met, Athena already knows me in a way no being has known me before.

And she seems as curious about me as I am about her. Slowly, she transfers her grip on me from the smaller, outer suckers at the tips of her arms to the larger, stronger ones nearer her head. I am now bent at a 90-degree angle, folded like a half-open book, as I stand on the little step stool. I realize what is happening: She is pulling me steadily into her tank.

How happily I would go with her, but alas, I would not fit. Her lair is beneath a rocky overhang, into which she can flow like water, but I cannot, constrained as I am by bones and joints. The water in her tank would come to chest height on me, if I were standing up; but the way she is pulling me, I would be upside down, headfirst in the water, and soon facing the limitations of my air-hungry lungs. I ask Scott if I should try to detach from her grip and he gently pulls us apart, her suckers making popping sounds like small plungers as my skin is released.

⚜

"Octopus?! Aren't they monsters?" my friend Jody Simpson asked me in alarm, as we hiked with our dogs the next day. "Weren't you scared?" Her question reflected less an ignorance of the natural world than a wide knowledge of Western culture.

A horror of giant octopuses and their kin, giant squid, has animated Western art forms from thirteenth-century Icelandic legends to twentieth-century American films. The massive "hafgufa" who "swallows men and ships and whales and everything it can reach" in the Old Icelandic saga *Orvar-Odds* is surely based on some kind of tentacled mollusk, and gave rise to the myth of the kraken. French sailors' reports of giant octopuses attacking their ship off the coast of Angola inspired one of the most lasting images of octopus in modern memory, one that is still tattooed on sailors' arms: Mollusk expert Pierre Denys de Montfort's iconic pen-and-wash drawing of 1801 shows a giant octopus rising from the ocean, its arms twisting in great loops all the way to the top of a schooner's three masts. He claimed the existence of at least two kinds of giant octopus, one of which, he concluded, was surely responsible for the disappearance of no fewer than ten British warships that mysteriously vanished one night in 1782. (To Montfort's public embarrassment, a survivor later revealed that they were really lost in a hurricane.)

In 1830, Alfred Tennyson published a sonnet about a monstrous octopus whose "Unnumber'd and enormous polypi / Winnow with giant arms the slumbering green." And of course an octopus was the nemesis-star of Jules Verne's 1870 French science-fiction novel, *Twenty Thousand Leagues Under the Sea*. Though the octopus becomes a giant squid in the 1954 film of the same name, the man who shot the underwater sequences for the original film in 1916, John Williamson, said this about the novel's original villain: "A man-eating shark, a giant poison-fanged moray, a murderous barracuda, appear harmless, innocent, friendly and even attractive when compared to the octopus. No words can adequately describe the sickening horror one feels when from some dark mysterious lair, the great lidless eyes of the octopus stare at one. . . . One's very soul seems to shrink beneath their gaze, and cold perspiration beads the brow."

Eager to defend the octopus against centuries of character assassination, I replied to my friend, "Monsters? Not at all!" Dictionary definitions of monster always mention the words *large, ugly*, and *frightening*. To me, Athena was as beautiful and benign as an angel. Even "large" is up for debate where octopuses are concerned. The largest species, the giant Pacific, isn't as big as it used to be. An octopus with an arm span of more than 150 feet may have once existed. But the largest octopus listed by *The Guinness Book of Records* weighed 300 pounds and boasted an arm span stretching 32 feet. In 1945, a much heavier octopus captured off the coast of Santa Barbara, California, was reported to weigh 402 pounds; disappointingly, a photo of this animal displayed with a man for size comparison suggests a radial span of only 20 to 22 feet. But even these modern giants hardly measure up to their close molluscan relative, the colossal squid. A recent specimen of this species, captured by a New Zealand boat fishing off Antarctica, weighed more than 1,000 pounds and stretched more than 30 feet long. These days, lovers of monsters lament that the biggest octopuses seem to have been captured more than half a century ago.

As I described Athena's grace, her gentleness, her apparent friendliness, Jody was skeptical. A huge, slimy cephalopod covered with suckers qualified as a monster in her book. "Well," I conceded, changing tacks, "being a monster is not necessarily a *bad* thing."

I've always harbored a fondness for monsters. Even as a child, I had rooted for Godzilla and King Kong instead of for the people trying to kill them. It had seemed to me that these monsters' irritation was perfectly reasonable. Nobody likes to be awakened from slumber by a nuclear explosion, so it was no wonder to me Godzilla was crabby; as for King Kong, few men would blame him for his attraction to pretty Fay Wray. (Though her screaming would have eventually put off anyone less patient than a gorilla.)

If you took the monsters' point of view, everything they did made perfect sense. The trick was learning to think like a monster.

After our embrace, Athena had floated back to her lair; I staggered down the three stairs of the step stool. I stood for a moment, almost dizzy, and caught my breath. The only word I could manage was "wow."

"The way she presented her head to you was unusual," said Scott. "I was surprised." He told me that the last two octopuses who lived here, Truman and, before him, George, would only offer their arms to a visitor—not the head.

Athena's behavior was particularly surprising given her personality. Truman and George were laid-back octopuses, but Athena had earned her name, that of the Greek goddess of war and strategy. She was a particularly feisty octopus: very active, and prone to excitement, which she showed by turning her skin bumpy and red.

Octopuses are highly individual. This is often reflected in the names keepers give them. At the Seattle Aquarium, one giant Pacific octopus was named Emily Dickinson because she was so shy that she spent her days hiding behind her tank's backdrop; the public almost never saw her. Eventually she was released into Puget Sound, where she had originally been caught. Another was named

Leisure Suit Larry—the minute a keeper peeled one of his questing arms off his or her body, two more would attach in its place. A third octopus earned the name Lucretia McEvil, because she constantly dismantled everything in her tank.

Octopuses realize that humans are individuals too. They like some people; they dislike others. And they behave differently toward those they know and trust. Though a bit leery of visitors, George had been relaxed and friendly with his keeper, senior aquarist Bill Murphy. Before I came, I had watched a video of the two of them together that the aquarium had posted on YouTube in 2007. George was floating at the top of the tank, gently tasting Bill with his suckers, as the tall, lanky aquarist bent down to pet and scratch him. "I consider him to be a friend," Bill told the cameraman as he ran his fingers over George's head, "because I've spent a lot of time interacting with him, taking care of him, and seeing him every day. Some people find them very creepy and slimy," he said, "but I enjoy it a lot. In some ways they're just like a dog. I actually pet his head or scratch his forehead. He loves it."

It doesn't take long for an octopus to figure out who his friends are. In one study, Seattle Aquarium biologist Roland Anderson exposed eight giant Pacific octopuses to two unfamiliar humans, dressed identically in blue aquarium uniforms. One person consistently fed a particular octopus, and another always touched it with a bristly stick. Within a week, at first sight of the people—looking up at them through the water, without even touching or tasting them—most of the octopuses moved toward the feeder and away from the irritator. Sometimes the octopus would aim its water-shooting funnel, the siphon near the side of the head with which an octopus jets through the sea, at the person who had touched it with the bristly stick.

Occasionally an octopus takes a dislike to a particular person. At the Seattle Aquarium, when one biologist would check on a normally friendly octopus each night, she would be greeted by a blast of painfully cold salt water shot from the funnel. The octopus hosed her and only her. Wild octopuses use their funnels not only for propulsion but also to repel things they don't like, just as you might use a snowblower to clear a sidewalk. Possibly the octopus was irritated by the night biologist's flashlight. One volunteer at the New England Aquarium always got this same treatment from Truman, who would shoot a soaking stream of salt water at her every time he saw her. Later, the volunteer left her position at the aquarium for college. Months later, she returned for a visit. Truman—who hadn't squirted anyone in the meantime—instantly soaked her again.

The idea of octopuses with thoughts, feelings, and personalities disturbs some scientists and philosophers. Only recently have many researchers accorded even chimpanzees, so closely related to humans we can share blood transfusions, the dignity of a mind. The idea set forth by French philosopher René Descartes in 1637, that only people think (and therefore, only people exist in the moral universe—"Je pense, donc je suis") is still so pervasive in modern science that even Jane Goodall, one of the most widely recognized scientists in the world, was too intimidated to publish some of her most intriguing observations of wild chimpanzees for twenty years. From her extensive studies at Gombe Stream Reserve in Tanzania, she had many times observed wild chimpanzees purposely deceiving one another, for example stifling a food cry to keep others from discovering some fruit. Her long delay in writing of it stemmed from a fear that other scientists would accuse her of anthropomorphizing—projecting "human" feelings onto—her study subjects, a cardinal sin in animal science. I have spoken with other researchers at Gombe who still haven't published some of their findings from the 1970s, fearing their scientific colleagues would never believe them.

"There's always an effort to minimize emotion and intelligence in other species," the New England Aquarium's director of public relations, Tony LaCasse, said after I met Athena. "The prejudice is particularly strong against fish and invertebrates," agreed Scott. We followed the ramp that spirals around the Giant Ocean Tank, affectionately known as the GOT, the three-story, 200,000-gallon re-creation of a Caribbean reef community that is the central pillar of the aquarium. Sharks, rays, turtles, and schools of tropical fish floated by like daydreams as we broke the scientific taboo and spoke of minds that many deny exist.

Scott remembered an octopus whose sneaky depredations rivaled those of Goodall's deceitful chimps. "There was a tank of special flounder about fifteen feet away from the octopus tank," he said. The fish were part of a study. But to the researchers' dismay, the flounder started disappearing, one by one. One day they caught the culprit red-handed. The octopus had been slipping out of her tank and eating the flounder! When the octopus was discovered, Scott said, "she gave a guilty, sideways look and slithered away."

Tony told me about Bimini, a large female nurse shark who once lived in the Giant Ocean Tank. One day the shark attacked one of the spotted eels in the tank and was swimming around with her victim's tail protruding from her mouth. "One of the divers who knew Bimini well wagged his finger at her, and then bopped her on the nose," Tony told me. In response, Bimini instantly regurgitated the eel. (Though the eel was whisked to the on-site veterinarian for emergency treatment, he unfortunately could not be saved.)

Once a similar thing had happened with our border collie, Sally. She had come upon a dead deer in the woods and was feeding on it. When I growled, "Drop it!" she actually vomited it up for me. I had always been proud of her obedience. But a shark?

The sharks don't eat all the fish in the tank, because they're well fed. "But sometimes they will eat or injure other animals for other reasons besides hunger," Scott told me. One day, a group of permits—long, thin, shiny fish whose dorsal fins are shaped like scythes—were thrashing around near the surface of the Giant Ocean Tank. "They were making a lot of noise and commotion," Tony said. One of the sand tiger sharks shot to the surface to attack the fish, biting their fins—but not killing or eating them. Apparently, the shark was just irritated. "This was a dominance bite, not a predator bite," Tony said.

To many, we spoke heresy. Skeptics are right to point out that it's easy to misunderstand animals, even those most like ourselves. Years ago, when I was visiting Birute Galdikas's research camp in Borneo, where ex-captive orangutans were learning to live in the wild, a new American volunteer, smitten with the shaggy orange apes, rushed up to an adult female to give her a hug. The female picked up the volunteer and slammed her against the ground. The woman didn't realize that the orangutan didn't feel like being grabbed by a stranger.

It's alluring to assume that animals feel as we do, especially when we want them to like us. A friend who works with elephants told me of a woman who called herself an animal communicator, who was visiting an aggressive elephant at a zoo. After her telepathic conversation with the elephant, the communicator told the keeper, "Oh, that elephant really likes me. He wants to put his head in my lap." What was most interesting about this interaction was the part the communicator may have gotten right: Elephants do sometimes put their heads in the laps of people. They do this to kill them. They crush people with their foreheads like you would grind out a cigarette butt with your shoe.

The early-twentieth-century Austrian-British philosopher Ludwig Wittgenstein once famously wrote, "If a lion could talk, we couldn't understand him." With an octopus, the opportunity for misunderstanding is greatly magnified. A lion is a mammal like us;

an octopus is put together completely differently, with three hearts, a brain that wraps around its throat, and a covering of slime instead of hair. Even their blood is a different color from ours; it's blue, because copper, not iron, carries its oxygen.

In his classic *The Outermost House*, American naturalist Henry Beston writes that animals "are not brethren, they are not underlings" but beings "gifted with extensions of the senses we have lost or never attained, living by voices we shall never hear." They are, he writes, "other nations, caught with ourselves in the net of life and time, fellow prisoners of the splendor and travail of the earth." To many people, an octopus is not just another nation; it's an alien from a distant and menacing galaxy.

But to me, Athena was more than an octopus. She was an individual—who I liked very much—and also, possibly, a portal. She was leading me to a new way of thinking about thinking, of imagining what other minds might be like. And she was enticing me to explore, in a way I never had before, my own planet—a world of mostly water, which I hardly knew.

❊

Back at home, I tried to replay my interaction with Athena in my mind. It was difficult. There was so much of her, everywhere. I could not keep track of her gelatinous body and its eight floaty, rubbery arms. I could not keep track of her continually changing color, shape, or texture. One moment, she'd be bright red and bumpy, and the next, she'd be smoother and veined with dark brown or white. Patches on different parts of her body would change color so fast—in less than a second—that by the time I registered the last change, she would be on to another. To borrow a phrase from songwriter John Denver, she filled up my senses.

Unconstrained by joints, her arms were constantly questing,

coiling, stretching, reaching, unfurling, all in different directions at once. Each arm seemed like a separate creature, with a mind of its own. In fact, this is almost literally true. Three fifths of octopuses' neurons are not in the brain but in the arms. If an arm is severed from an octopus's body, the arm will often carry on as if nothing has happened for several hours. One presumes the severed arm might continue hunting and perhaps even catching prey—only to pass it back toward a mouth to which the arm is, sadly, no longer attached.

Just one of Athena's suckers was enough to seize my complete concentration—and she had 1,600 of them. Each was busily multi-tasking: sucking, tasting, grabbing, holding, plucking, releasing. Each arm on a giant Pacific octopus has two rows of suckers, the smallest at the tips, the largest (three inches across on a big male, perhaps two on Athena) about a third of the way to the mouth. Each sucker has two chambers. The outer one is shaped like a broad suction cup, with hundreds of fine radial ridges stretching to the rim. The inner chamber is a little hole in the center of the sucker, which creates the suction force. The whole structure can bend to fit the contours of whatever the sucker is grasping. Each sucker can even fold to create a pincer grip, like your thumb and forefinger can. Each is operated by individual nerves that the octopus controls voluntarily and independently. And each sucker is fantastically strong. James Wood, webmaster of the long-running biological website The Cephalopod Page, has calculated that a 2.5-inch-diameter sucker can lift 35 pounds of weight. If all the suckers were that size, the octopus would have a sucking capacity of 56,000 pounds. Another scientist calculated that to break the hold of the much smaller common octopus would demand a quarter ton of force. "Divers," Wood said, "should be very careful."

Athena's suction had been tender with my skin. Since I was not afraid, I had not resisted her pull. This was fortunate, I learned when I later spoke with her keeper, Bill, on the phone, setting up my next visit.

"A lot of people are freaked out by them," he told me. "When visitors come, we always have someone there to help in case the person freaks out. Keeping the octopus in the tank is the main goal. We can't guarantee what they'll do. With Athena, I've had four of her arms on me, and you peel them off and then the other four arms are on."

"I think we've all been on dates like that," I observed.

While Athena was tasting my arms and hands, she had made a point of looking into my face. I was impressed that she even recognized a face so unlike her own, and wondered whether Athena might like to taste my face as well as look at it. I asked Bill if that was ever allowed. "No," he said emphatically, "we don't let them near the face." Why? Could she pull out an eye? "Yes," Bill said, "she could." Bill has gotten into futile tugs-of-war with octopuses who have grabbed the handles of cleaning brushes. "The octopus always wins. You have to know what you're doing," he said. "You cannot let her go near your face."

"I felt as if she wanted to pull me into the tank," I told him.

"She could pull you into the tank, yes," he said. "She will try."

I was eager to give her another chance. We set a date for a Tuesday, when both Bill and his most experienced octopus volunteer, Wilson Menashi, would be there. Scott, and now Bill, told me the same thing about Wilson: "He has a real way with octopuses."

Wilson is a former engineer and inventor with the Arthur D. Little Corp. with many patents to his name. Among his other accomplishments is having brought cubic zirconia to market as an imitation diamond. (It had been artificially produced by the French, but they didn't know what to do with it.) At the aquarium, Wilson had been tasked with an important mission: designing interesting toys to keep the intelligent octopus occupied. "If they have nothing to do, they become bored," Bill explained. And boring your octopus is not only cruel; it's a hazard. I knew from living with two border collies and a 750-pound pet pig that to allow a smart animal to become bored is to court disaster. They will invariably come up with something creative to do with their time that you don't want them to do, as the Seattle Aquarium had discovered with Lucretia McEvil. In Santa Monica, a small California two-spot octopus, only perhaps eight inches long, managed to flood the aquarium's offices with hundreds of gallons of water by experimenting with a valve in her tank, causing thousands of dollars' worth of damage by ruining the brand-new, ecologically designed floors.

Another danger of boredom is that your octopus may try to go someplace more interesting. They are Houdini-like in their ability to escape their enclosures. L. R. Brightwell of the Marine Biological Station in Plymouth, UK, once encountered an octopus crawling down the stairs at two thirty in the morning. It had escaped from its tank in the station's laboratory. While on a trawler in the English Channel, an octopus who had been caught and left on deck somehow managed to slither from the deck, down the companionway, to the cabin. Hours later, it was found hiding in a teapot. Another octopus, held in a small private aquarium in Bermuda, pushed off the lid from its tank, slid to the floor, crawled off a veranda, and headed home to the sea. The animal had traveled about 100 feet before it collapsed on the lawn, where it was attacked by a horde of ants and died.

Perhaps an even more surprising case was reported in June 2012, when a security officer at California's Monterey Bay Aquarium found a banana peel on the floor in front of the Shale Reef exhibit at 3 a.m. On closer inspection, the banana peel turned out to be a healthy, fist-size red octopus. The security officer followed the wet slime trail and replaced the octopus in the exhibit it had come from.

But here's the shocking part: The aquarium didn't know it had a red octopus living in its Shale Reef exhibit. Apparently the octopus hitchhiked there as a juvenile, attached to a rock or sponge added to the exhibit, and grew up at the aquarium without anyone knowing it was there.

To avert disaster, aquarium staff carefully design escape-proof lids to their octopus tanks and try to invent ways to keep their octopuses occupied. In 2007, the Cleveland Metroparks Zoo put together an enrichment handbook for octopus, filled with ideas of how to keep these smart creatures entertained. Some aquariums hide food inside a Mr. Potato Head and let the octopus dismantle the toy. Others offer Legos. Oregon State University's Hatfield Marine Science Center has devised a contraption that allows an octopus to create art by moving levers that release paint onto a canvas—which is then auctioned to generate funds to maintain the octopus tank.

At the Seattle Aquarium, Sammy the giant Pacific octopus enjoyed playing with a baseball-size plastic ball that could be screwed together by twisting the two halves. A staffer put food inside the ball but later was surprised to find that not only had the octopus opened the ball, it had *screwed it back together* when it was done. Another toy was constructed from the plastic tubing through which pet gerbils like to tunnel. Rather than probe into the tunnel with his arms, which was what the aquarists had expected, Sammy liked to unscrew the pieces—and when he was done, he handed them off to his tank mate, an anemone. The anemone, who, like all of its kind, was brainless, held on to the pieces with its tentacles for a while, bringing them to its mouth, and finally spat them out.

But Wilson was ahead of the curve. Long before the first octopus-enrichment handbook was published, many octopuses ago, he set out to create a safe toy worthy of an octopus's intellect.

Working at his lab at Arthur D. Little Corp., Wilson devised a series of three clear Plexiglas cubes with different locks. The smallest of the three has a sliding latch that twists to lock down, like the bolt on a horse's stall. You can put a live crab—a favorite food—inside and leave the lid unlocked. The octopus will lift the lid. When you lock the lid, invariably the octopus will figure out how to open it. Then it's time to deploy the second cube. This one has a latch that slides counterclockwise to catch on a bracket. You put the crab in the first box and then lock it inside the second box. The octopus will figure it out. And finally, there's a third cube. This one has two different latches: a bolt that slides into position to lock down, and a second one with a lever arm, sealing the lid much like an old-fashioned canning jar closes. Bill told me that once the octopus "gets it," the animal can open all four locks in three or four minutes.

I was looking forward to meeting Bill and Wilson, and was hungry to hear what they had to tell me. But even more, I longed to see Athena again and to learn how she behaved among people she knew. And I wondered: Would she recognize me?

※

Bill meets me in the aquarium lobby. He's thirty-two years old, six foot five, slender, and strong, with short brown hair and a smile that takes over his whole face, crinkling the edges of his eyes. Tentacles creep down from under the right sleeve of his green aquarium shirt—the tattoo of a Portuguese man-of-war, a stinging jellyfish, with an azure sail. We walk up the staircase to the aquarium café, and then take the Employees Only stairway to the Cold Marine Gallery, which Bill runs. He's in charge of 15,000 animals here, from invertebrates like Athena and the sea stars and anemones, to giant lobsters and endangered turtles and the strange, ancient chimera, or ghost shark—a deepwater species with grinding, instead of sharp, teeth, whose cartilaginous kind branched off from the shark lineage 400 million years ago. Bill knows each of his charges personally; he has known many of them since they were born (or hatched, or budded) under his care; many others, he collected on expeditions to the chilly waters of Maine and the Pacific Northwest.

Wilson is already here. He's a much smaller man than Bill, trim and quiet, with a dark moustache, the hairline of a grandfather of nearly grown grandchildren, and a Middle Eastern accent I can't quite place. He looks much younger than his seventy-eight years.

It's nearly 11 a.m., Athena's feeding time. A dish of silvery, five-inch capelin awaits her, sitting on the lid of an adjoining tank. We don't want to keep her waiting.

The men heave the heavy tank top up and attach it to an overhead hook to keep it propped open. The lid is covered in fine mesh and precisely contoured to fit the elaborate curves of the tank's outlines, a precaution perfected over the course of many octopuses, to prevent escape. Bill leaves me with Wilson to attend to other chores in his gallery. Wilson mounts the short movable stair and leans over the tank.

Athena rises up from her lair like steam from a pot. She's coming to Wilson so quickly it takes my breath away—much faster than she had come to see me earlier.

"She knows me," Wilson states simply. He reaches into the cold water to greet her.

Athena's white suckers arch from the water to grasp Wilson's hands and forearms. She looks at him with her silvery eye, then surprises me: She flips over, like a puppy showing its belly. Wilson hands a fish to the center suckers on one of her front arms. The food heads toward her mouth like on a conveyor belt as she passes it from sucker to sucker. I'm eager to see inside her mouth, glimpse her beak. But I am disappointed. The fish disappears like the stairs at the end of an escalator. Wilson says he's never known an octopus to show its beak.

Only now do I notice that a large orange sunflower sea star is moving toward Wilson's hand. With more than twenty limbs, called "rays," befitting a star, and an arm span of more than two feet, it's edging toward us on 15,000 tube feet. Like all sea stars, this largest of all the species has no eyes, no face, and no brain. (As an embryo, the sea star starts to grow one, but apparently thinks better of it and instead forms a neural net around the mouth.)

"He wants a fish too," Wilson says. (This sea star is, in fact, male, as became evident when he released his sperm one day, clouding the tank.) Wilson hands him a capelin with the same easy motion with which one might pass the butter dish to a guest at the dinner table.

How can a brainless animal "want" anything—much less communicate its desires to another species? Perhaps Athena knows. To her, the sea star may be a distinct individual, a neighbor whose habits and quirks she recognizes and anticipates. At the Hatfield Marine Science Center's Visitor Center, when the octopus was done playing with Mr. Potato Head, the sea star would take the eyes and carry them around between two of his arms. ("He looked really cute," Kristen Simmons, who invented the painting apparatus for the octopus, told me.) She described their sea star as "inquisitive" and told me that whenever the octopus gets a new toy, the sea star "tries to take it away from him—which I find amazing." If a staffer moved a toy away from this sea star, the animal would hurry to retrieve it.

I wonder: Can a brainless animal feel curiosity? Does it want to play? Or does it only "want" toys or food the way a plant "wants" the sun? Does a sea star experience consciousness? If it does, what does consciousness feel like to a sea star?

Clearly, I have entered a world I cannot judge by the rules I have learned on land among vertebrates. The sea star begins to dissolve the fish before our eyes, the capelin melting away as though viewed via time-lapse photography. The sea star can extrude its stomach outside the mouth to digest prey, which is usually sea urchins, snails, sea cucumbers, and other sea stars.

The sea star sated, Wilson turns back to Athena and feeds her the rest of the fish. He hands her one fish after another, three more in all. He deposits each in the suckers of a different arm. I watch in astonishment as the octopus conveys each fish along her suckers, toward her mouth. It seems to take a long time before each fish reaches its destination. Why doesn't she just flex the arm and place the fish directly in her mouth? Then it occurs to me: Perhaps it's for the same reason we lick an ice cream cone instead of shoving it past the tongue down the throat. Taste is pleasurable, and it's pleasurable because it's useful: this is how we know what is good and safe to eat and what is inedible. An octopus does the same with its suckers.

Once Athena finishes eating her fish, she plays gently with Wilson's hands and forearms. Occasionally the tendril-like tip of an arm curls up to his elbow, but almost lazily; mostly her arms twist weightlessly in the water, her suckers gently kissing his skin. With me, before, her suction had felt exploratory, insistent. But with Wilson she is completely relaxed. As I look at the man and the octopus touching each other, they remind me of a happy older couple, many years into a loving marriage, tenderly holding hands.

I put my hands in the water with Wilson's and touch one of Athena's unoccupied arms. I slowly stroke some of her suckers. They fold to fit the contours of my skin and latch on. I can't tell if she recognizes me. Though I am sure she can taste I am a different person, Athena seems to consider me a part of Wilson, the way a person might behave toward a companion that a trusted friend has brought along. Athena latches onto my skin slowly, languidly, the same way she did greeting Wilson. I lean over to glimpse her pearly eye, and she pulls her head to the surface to look me in the face.

"She has eyelids like a person does," says Wilson. He gently passes his hand over her eyes, causing her to slowly wink. She doesn't recoil or move away. The fish are gone; she is staying near the surface for the company.

"She's a very gentle octopus," Wilson says, almost dreamily, "very gentle. . . ."

Has working with octopuses made him gentler or more compassionate? Wilson pauses. "I don't have the language to answer that question," he says. Wilson was born along the Caspian Sea, in Iran, near Russia, and spoke Arabic before he learned English as a small child because his parents were from Iraq. He doesn't mean that he lacks the English skills to answer. He means that he hasn't thought of this before. "I've always liked toddlers and kids," he says. "I can relate to them. This is . . . similar."

As with a child, to commune with Athena demands a level of openness and intuition greater than that used in the usual discourse between adult humans of a common culture. But Wilson doesn't equate this strong, smart, wild-caught adult octopus to a baby human—unfinished, incomplete, not quite fully developed. Athena is, in the words of the late, great Canadian storyteller Farley Mowat, "more-than-human," a being who doesn't need us to bring her to completion. The wonder is that she will allow us to be part of her world.

"Don't you feel honored?" I ask Wilson.

"Yes," he says emphatically. "Yes."

Bill, rejoining us from his errands, leans his tall frame over the tank and reaches in to stroke Athena's head.

"It's a rare pleasure," Bill says. "Not everyone can do this."

※

How long did we stay with Athena? It's impossible to say. Of course, we had removed our watches before plunging our arms into the water. Once we did, we entered what we called Octopus Time. Feelings of awe are known to expand the human experience of time availability. So does "flow," the state of being fully immersed in focus, involvement, and enjoyment. Meditation and prayer, too, alter time perception.

And there is another way we alter our experience of time. We as well as other animals can mimic another's emotional state. This involves mirror neurons—a type of brain cell that responds equally whether we're watching another perform an action, or whether we're performing that action ourselves. If you are with, for example, a calm, deliberate person, your own perception of time may begin to match his. Perhaps, as we stroked her in the water, we entered into Athena's experience of time—liquid, slippery, and ancient, flowing at a different pace than any clock. I could stay here forever, filling my senses with Athena's strangeness and beauty, talking with my new friends.

Except our hands froze—so red and stiff that we could not move our fingers. Taking our hands out of Athena's tank felt like breaking a spell. I was suddenly desperately uncomfortable, awkward, and incompetent. Even after rinsing my red skin with hot water for nearly a minute, I was so cold I still couldn't pick up the pen in my purse, much less write in my notebook. It was as if I had trouble returning to the person, the writer, I was before.

⁂

"Guinevere was my first," Bill tells us, "so she's my favorite." Bill, Scott, Wilson, and I have gone to a nearby sushi place for lunch. I think it an odd choice, but perhaps not; we have just been watching Athena eat raw fish, after all. No one orders octopus. I get California rolls.

"The first two minutes you interacted with her, Guinevere was all over you," Bill continued. But then she'd calm down, staying close by and exploring Bill's arms gently with her suckers.

Guinevere was also the first and only octopus who ever bit Bill. She didn't envenomate him, and the bite didn't leave a scar. Still, he admitted, "I don't want it to happen again." It was like a bite from a parrot, he said. A parrot can exert 600 pounds of pressure per square inch with its beak, so this was not a small thing, but Bill shrugged it off. As if to clear Guinevere's reputation, he added, "It was not a huge bite."

It had happened early in their relationship. And besides, he added gallantly, it had been his fault. He had let his hand get too close to her mouth. "She was curious: 'Can I eat you?'"

The guys tell me about the other octopuses they've known.

"George was really good," Bill said. "He was pretty calm. He was a pretty good octopus—not feisty. The feisty ones are the ones that the first ten minutes you spend pulling arms off you. They're constantly grabbing at you. George would come over, crawl on your arm, eat, then move on. Sometimes we'd hang out for an hour together.

"George died while I was on vacation," he continued. Octopuses live fast and die young: Giant Pacific octopuses are probably among the longest-lived of the species, and they usually live only about three or four years. And by the time they arrive at the aquarium, they are usually at least a year old, sometimes more. "I had no idea George was about to die," Bill said. "Usually they change in body and behavior and coloration. They don't stay as red. They're whitish all the time. The intensity isn't there. They're less playful. It's like old age in people. Sometimes they get age spots, white patches on their skin that seem to be sloughing off."

"That must be so hard," I said to Bill. He shrugged. This is, after all, part of the job. But on my first visit, Scott had said, about Bill and his octopuses: "They're like his babies. When one passes away, it's a loss. That's an animal he's loved and cared about every day for years."

George's successor, Truman, arrived while Bill was away. "He was one of the most active octopuses from the start. Truman," he said, "was an opportunist."

Different octopuses had different approaches to opening Wilson's boxes. Each learned fairly quickly how to open the locks. Bill would start with the smallest box and present it to the octopus once a week for about a month. At two months they'd try the second box. They mastered it in two to three weekly tries. The third box, with its two different locks, might take five or six tries. But even though everyone mastered the locks, on occasion each octopus, depending on personality, might employ a different strategy.

Calm George always opened the locks methodically. But Guinevere was impetuous. One day, the live crab inside so excited her that she squeezed the second-largest box hard enough to crack it. Later, when Truman was introduced to the boxes, he seemed to enjoy opening them. But one day Bill gave him a special treat, putting two live crabs inside the smallest box. When the two crabs started fighting, Truman became too excited to bother with the locks. He poured his seven-foot-long body through the two-by-six-inch crack Guinevere had made. Visitors to his exhibit found the giant octopus, suckers flattened and facing out, squeezed into the tiny space between the walls of the fourteen-cubic-inch middle box and the six-cubic-inch one inside it. Truman never did open the small box. Probably he was too cramped. But when he finally emerged from his cube, Bill fed him both crabs anyway.

Because octopuses can squeeze into such small spaces, aquarists have had some frightening moments. George scared Bill nearly to death one day, when he'd hidden underneath a big rock, and Bill couldn't find him even after a long, frantic search. "I thought he'd escaped," Bill said.

"Any hole, they're going to go right through it," Wilson agreed.

More than a decade earlier, Scott had known a dwarf Caribbean octopus who lived in one of the smaller display tanks known as jewel cases at the aquarium. One day Scott came in to work to find the tank overflowing onto the floor, and the octopus nowhere in sight. He found that the animal had oozed behind the background of its exhibit and wedged itself into the half-inch-diameter pipe that re-circulated the water. What to do?

"I remembered having watched this National Geographic show as a kid," he said. It had showed fishermen in Greece pulling up am-phora pots they had set for octopuses. After hunting all night, the octopuses thought they had found safe dens there, only to be hauled up by fishermen who wanted to eat them. Naturally they didn't want to come out of the pots, and the fishermen didn't want to break their vessels, so they had poured fresh water into the pots, and the octo-puses came rushing out. So Scott did the same with the dwarf Ca-ribbean octopus—and it worked.

He employed the same method years later with a misbehav-ing giant Pacific, so long ago Scott doesn't remember the octopus's name, but he vividly recalls the incident. When Scott lifted the lid to the tank to feed the animal, the octopus attached to his hands and arms. When he'd peel one arm off, he'd find two more stuck to him. "The octopus wouldn't go back inside the tank, and I had to move on," he said. "I had things to do." So he reached to the sink across from the tank, filled a pitcher with fresh water, and poured it on the octopus. She instantly recoiled. "I'm thinking: I outwitted the octo-pus!" he said. Scott was rather proud of himself.

But the octopus was incensed. "She got scarlet red and really thorny. It was a heated moment. What I didn't notice," he said, "was she was blowing herself up." She siphoned up a massive load of water "and gushed a major surge of salt water onto my face!" As he stood there dripping, Scott noticed "the octopus had the same look on her face as I must have had on mine when I thought I'd outwitted her."

⁂

A few weeks later, I visited Athena for a third time. Bill and Wilson were both absent, so Scott opened the top of her tank for me. Athena had been resting in her usual lair, in a corner under a rock overhang, but she floated quickly to the top and hung before me, upside down.

I was disappointed at first that she didn't present her head or look at me. Was she less curious about me now? Had she glimpsed me coyly, like a woman behind a veil, peeking over the webbing between her arms, when I hadn't noticed? Did she rely on her suckers to tell her, even before she had touched me, who I was? If she did recognize me, though, why did she not approach me in the same way as before? Why was she hanging before me like an opened umbrella, upside down?

And then I realized what she wanted. She was asking me for food.

Scott asked around, and learned that Athena, who doesn't need to eat daily, hadn't been fed for a couple of days. And then he allowed me the privilege of handing her a capelin. I handed a fish to one of her large suckers. Athena began to convey the fish toward her mouth. But first she covered it with two of her other arms, enveloping it with many more suckers, as if she were licking her fingers, savoring the meal.

Once she had eaten, I reached deeper into the water. Now she let me pet her. As I stroked her head and mantle, I marveled again at her softness and texture: Her skin had gathered into little bumps and ridges. I reached for the webbing between her arms, which was as delicate as gossamer, and so thin I could see bubbles beneath it, as sometimes happens with a swimsuit. And yet, this body, so unlike my own, was responding to my touch like a dog's or a cat's or a child's. Even though her skin can change color and taste flavors, it, like mine, relaxes into a caress. And though her mouth is between her arms, and her saliva dissolves flesh, she, like me, clearly enjoys a good meal when she's hungry. I felt as if I had understood something very basic about her at that moment. I don't know what it's like to change color or shoot ink, but I do know the joys of gentle touch and of eating food when hungry. I know what it feels like to be happy. Athena was happy.

I was too. As I drove home to New Hampshire, my happiness swelled to elation. Now that I have fed her, I thought, surely she will remember me next time, if she doesn't already.

❧

A week later, I was shocked to receive this e-mail from Scott:

"Sorry to write with some sad news. Athena appears to be in her final days, or even hours." Less than an hour later he wrote again that she was gone.

To my surprise, I broke down in tears.

Why such sorrow? I don't cry often. I would have been sad, but probably would not have wept, over a person I had met only thrice, with whom I had spent, in total, less than two hours. I had no idea whether I meant anything at all to Athena, and even if I had, it was surely little. I was not, like Wilson and Bill, Athena's special friend. But she meant a great deal to me. She was, like Bill's Guinevere, "my first." We had hardly known each other, but she had given me a glimpse into a kind of mind I had never known before.

And that was part of the tragedy: I had just started to know her. I was mourning the relationship that could have blossomed but didn't have a chance to grow.

"What is it like to be a bat?" the American philosopher Thomas Nagel famously asked in his 1974 essay on the subjective nature of consciousness. Many philosophers might argue that to be a bat is not "like" anything—for, according to some, animals do not experience consciousness. A sense of self is an important component of consciousness, one that a number of philosophers and researchers claim humans have but animals don't. If animals were conscious, according to one book, written by a Tufts University professor, dogs would untangle their leashes from poles and dolphins would leap out of tuna nets. (That author clearly doesn't read Dear Abby. Why don't those women leave their abusive husbands? Why won't that couple just stop visiting the rude in-laws?)

Nagel concluded, like Wittgenstein before him, that it is impossible to know what it is like to be a bat. After all, a bat sees much of its world using echolocation, a sense we do not possess and can hardly imagine. How much further from our reach is the mind of an octopus?

Yet still I wondered: What is it like to be an octopus?

Isn't this what we want to know about those whom we care about? What is it like, we wonder at each meeting, in shared meals and secrets and silences, with each touch and glance, to be you?

"There is a young pup octopus headed to Boston from the Pacific Northwest," Scott wrote me days after Athena had died. "Come shake hands (x8) when you can."

At Scott's invitation, I set out to cross a chasm of half a billion years of evolution. I set out to make an octopus my friend.

CHAPTER TWO

Octavia

This Shouldn't Be Happening: Tasting Pain, Seeing Dreams

Hello, beauty!" I greeted the new octopus as I perched next to Wilson on the step stool, leaning over the top of her tank. Even though I couldn't see her now, I knew she was beautiful, because I had seen her moments before from the public side. I couldn't wait to meet her. She was much smaller and more delicate than Athena, with a head the size of a large clementine. All of her skin was dark brown and thorny, and she'd been plastered by her white suckers to the front of the glass. Her largest suckers were less than an inch in diameter; the smallest, tinier than pencil points. Her silvery eye had peeked out from behind the hedge of her arms.

"What's her name?" I called out to Bill, behind us, who was adjusting the filter on a tank temporarily housing a grunt sculpin, a bug-eyed fish with a face like a Boston terrier.

"Octavia," he called back, over the din of the pumps and filters. A little girl visiting the aquarium had come up with the name and Bill thought it was a good one.

Octavia had come from British Columbia, where she had been caught wild and shipped to the aquarium at greater expense than she cost to buy, via Federal Express. I had waited impatiently for several weeks before coming to meet her, to give her a chance to settle in. With me today was my friend Liz Thomas, the author and anthro-

pologist, who is as drawn to those whom Canadian author Farley
Mowat calls "the Others" as I am. As a teenager in the 1950s, she had
lived with her parents among the Bushmen in Namibia, which she
wrote about in her first best seller, *The Harmless People*; she spent
the next six decades researching and writing nonfiction books about
lions, elephants, tigers, deer, wolves, and dogs, as well as two Paleo-
lithic novels. She wanted to touch an octopus too.

Wilson tried to entice Octavia to us with food. At the end of a
pair of long tongs, he offered her a squid, an octopus relative. She
didn't even extend an arm.

"Come over and see us, you pretty little thing!" Pleading with an
invertebrate (let alone one without ears) might seem a crazy thing
to do, but I couldn't help but speak to her, as I would to a dog or a
person. Wilson waved the squid so that its eight arms and two feeding tentacles floated in an almost lifelike fashion, spreading its taste
through the water. Octavia surely sensed it with her skin and suckers. She surely saw it too. But she wanted nothing to do with it—or
with us.

"Let's try again later," said Wilson. "She might change her
mind."

While Wilson attended to chores with Bill, Liz and I visited
the spiraling walkway that wraps around the Giant Ocean Tank.
At the lower levels, electric-blue chromis and flamboyant yellowtail
damselfish darted in and out of fiberglass corals; yellowtail snapper
swam by in packs, like groups of teens at a mall. Higher up, rays flew
by on cartilaginous wings, while their relatives the sharks cruised
sinuously and purposefully, as if on urgent errands. Huge turtles
oared the water with scaly flippers. Everyone's favorite, Myrtle, a
green sea turtle who weighs 550 pounds, is known as the Queen of
the Giant Ocean Tank. Myrtle has been here since the aquarium was
a year old, and she dominates even the sharks, stealing squid right

out of their toothy mouths. Generations of children have grown up knowing this personable and fearless turtle, who swims right up to the glass to look you in the face, who loves it when divers scratch her back (turtles have nerve endings in the carapace), and who has been known to fall asleep in the lap of one of her favorite aquarists, Sherrie Floyd Cutter, as Sherrie pats her head. Myrtle even has her own Facebook page, which on any given day may garner well over one thousand "likes."

Myrtle is thought to be about eighty (and if so, she could live long enough to see today's toddlers bring their own children to the aquarium). But even at her advanced age, not long ago, Myrtle was part of a landmark study that proved that reptiles—even old reptiles—can learn new tricks. Myrtle was presented with three small platforms: Two had speakers, and the middle one had a light box. If the light in the light box went on, she was to touch the box with a flipper. But if the light went on with a tone, she had to decide which speaker was making the tone and touch that platform instead. This was more than a trick. It was a complex task, because it involved more than responding to one request or command. It demanded that Myrtle make a decision.

"Think of all the things that turtle has seen and learned in eighty years," Liz said as Myrtle winged her way past us. Most people think that turtles are slow, but green sea turtles can actually swim 20 miles an hour when they're in a hurry, and Myrtle was heading to the top of the tank, where a diver had appeared with food. "Brussels sprouts are Myrtle's favorite vegetable," the diver was telling the crowd. ("Ew! Brussels sprouts!" a little girl said to her older brother.) But food isn't all that's on this turtle's mind. "She seems genuinely interested in things we're doing," Sherrie says, "even when we don't have food. Almost to the point of being nosy about everything that goes on in the tank. Whenever we're at the platform, she's hanging all over us, trying to see what's on the platform, around the platform, and I'm constantly pushing her away." During promotional and film shoots at night in the GOT, the aquarium has to deploy one diver for the express purpose of distracting Myrtle, so she won't get into the shot. And even this ploy is only effective for about ninety seconds before the turtle swims to where the action is.

When we went back upstairs to Cold Marine for another try with Octavia, she still wasn't interested. I tried to fathom her shyness. Why wouldn't she come over to see us?

"Everyone is different," Wilson reminded us. "They have different personalities. Even lobsters have different personalities. You stick around here long enough and you'll see."

Already it was clear that Octavia was quite different from Athena. Octavia's situation was unusual, Wilson explained. Athena had died suddenly and unexpectedly. Usually, octopuses show signs of aging—develop white spots, stop eating, grow thin—and the aquarium orders a new one. The young pup grows up behind the scenes, and so is acclimated to people by the time the old octopus dies and the exhibit tank is free. "Those who grow up at the aquarium are usually friendly," Wilson said. "They are the most playful. They're like puppies and kittens."

But the aquarium didn't have time for a younger octopus to grow up—they needed a display octopus immediately. "An aquarium without an octopus," as the Victorian naturalist Henry Lee of Brighton, UK, wrote in 1875, "is like a plum pudding without plums." So Bill had ordered from his supplier a new octopus big enough to impress the public.

Octavia might be two and a half years old already. Because she had grown up in the wild (giant Pacific octopuses are not raised in captivity, Bill explained, and their ocean populations are thought to be healthy), she hadn't yet warmed to human company.

Wilson tried one last time. He held out the squid to Octavia on the grabber, and a single arm came floating tentatively over.

"Liz! *You* touch her!" I cried, sensing that the opportunity for interaction might be fleeting. My friend mounted the three little stairs to the top of the tank and extended her index finger to the tendril-like tip of Octavia's arm. The scene reminded me of the ceiling of the Sistine Chapel, showing Adam extending his hand to God in heaven.

The encounter lasted just a moment. Liz felt the slender, slippery back of the tip of Octavia's arm, and Octavia twisted it over to taste Liz gingerly with her elfin suckers.

Both of them instantly withdrew in alarm.

Liz is not frightened by animals, or by anything else, for that matter. The first day we met, nearly thirty years ago, I introduced her to one of our ferrets, who immediately sank his pointy teeth into her hand, drawing blood.

"Sorry," I said.

"I don't mind *at all*," Liz told me, and she meant it. She's spent days alone with wolves in the Arctic and been stalked by a wild leopard in Uganda. In Namibia, when a hyena, a species that has been known to bite off the noses of sleeping people, thrust her head into Liz's tent, my friend's only response was to ask the bone-crushing carnivore, "What is it?" as if her mom had appeared at the door to her room. But Liz said that Octavia's touch had been "viscerally surprising." Her response was atavistic. Liz couldn't help but spring backward.

And what had alarmed Octavia about Liz? I couldn't be sure, of course, but Liz is a pack-a-day smoker. I wondered whether Octavia's exquisitely tuned senses, with 10,000 chemoreceptors on each sucker, had tasted the nicotine on Liz's skin or even in her blood. Nicotine is a known insect repellent, and toxic to many other invertebrates. Liz's finger may have tasted, to Octavia, just plain icky. I hoped this wouldn't make her think we all tasted bad.

❧

On my second visit, I waved a dead squid back and forth in the cold water till my right hand cramped and I could no longer move it. I switched to my left hand till it froze, too. Octavia stayed way over on the opposite side of the tank. She didn't even extend an arm.

This was a Friday, and Wilson wasn't there. I went downstairs to get a better view of Octavia. She was thorny and dark, and almost invisible in the dim light of her rocky lair. Because giant Pacific octopuses are, like most species of octopus, nocturnal, the tank isn't brightly illuminated, and conveys quiet mystery. Her only tank mates, the sunflower sea star, about forty rose anemones, and two types of starfish, a bat star and a leather star, sat affixed to their stations. Anchored by his thousands of tube feet, the sea star had taken up what seemed to be his usual position, opposite the octopus. This is a good spot from which to snag a fish should an aquarist open the tank. Sunflower sea stars can move quickly for starfish, three feet per minute if in a hurry, but even with no brain, he seemed to know he wasn't as fast as an octopus.

The anemones' tentacles swayed in the water like the petals of flowers in a breeze. In fact anemones look like plants but are actually predatory invertebrate animals, like Octavia and starfish—but more closely related to corals and jellyfish. They attach themselves to the substrate with sticky feet, harpooning small fish and shrimp with organs called nematocysts and injecting their prey with stinging venom.

Octavia appears to be sharing her tank with two lugubrious-looking wolf eels and a number of species of big rockfish with spiny, often poisonous dorsal fins—though she's really not.

All of them would normally be found together in the wild in their native waters in the Pacific Northwest; but here the octopus is separated by a pane of glass from the eels and rockfish so they don't eat each other. The wolf eels' tank has brighter lighting, so the exhibit gives you the feeling that you're peeking in on a wild octopus in her lair, and out from there into the open ocean.

I waited for Octavia to move; for the tip of an arm to twitch, for her one visible eye to swivel and meet ours, for a change in her color. She remained immobile, her arms balled up, her head protected. I could not even see the white insides of her gills flashing as she breathed. She may have been watching us, but her slit-pupil eye betrayed nothing.

Eager to show me something moving, Scott took me over to see a favorite tank in his Freshwater Gallery: the electric eel exhibit. He's quite proud of it, and rightly so. Even though an electric eel is not colorful or cute or especially pretty ("You might find more attractive things in your toilet," Scott admitted), this tank is one of the most popular in the aquarium, with a gorgeous, naturalistic display. Scott has traveled many times to the Amazon, where he cofounded the nonprofit Project Piaba, supporting sustainable fisheries for aquarium fish. He knows what electric eel habitat looks like, so the tank is lush with living, native Amazon water plants. The electric eel loves to hide among the leaves. But this posed a problem for the viewing public. "They could never find the animal in the eel tank," he said. He realized what he had to do: *train the electric eel.*

It only took Scott a matter of weeks to teach the eel a completely unnatural behavior: to emerge from his comfortable hiding place in the vegetation, out into the open where visitors could see him. For this purpose, Scott invented a device he named the Worm Deployer.

Hanging above the eel's tank is a rotating electric fan, to which, attached by a Barrel of Monkeys plastic monkey, hangs an ordinary kitchen funnel. Staff periodically drop live earthworms into the funnel, which fall slowly into the water along the fan's arc, right in front of the public. "The eel never knows when manna might drop from heaven," Scott explained, "so he learned to hang out there, just in case." The only downside of his invention was that the exhibit used to have two electric eels, and the Worm Deployer caused them to fight. Now one of the eels has been exiled to a large tank near Scott's desk.

The Worm Deployer has many uses. Sometimes Scott uses it to manipulate the public. On busy days when visitors are jammed up in one particular area of the aquarium, Freshwater staff can usually break up the clot with a handful of worms, instantly drawing a crowd to the electric eel tank. The exhibit has another feature to keep the public riveted: A voltmeter picks up the fish's electric pulse. A light, actually powered by the eel's electricity, flashes across a panel built on top of the tank to show when the eel is hunting or stunning prey, and this quickly attracts attention.

On this morning, Scott and I had the eel tank to ourselves. Even though Scott had just fed some worms into the Deployer, the three-foot, reddish-brown eel was immobile. I wondered if he was just watchfully waiting. "Look at his face," Scott said. "No, that eel is catching some serious Zs." A worm dropped right near his head, and still the fish didn't move. The eel was fast asleep.

Then suddenly, we saw the voltmeter flash.

"What's going on?" I asked Scott. "I thought the eel was asleep."

"He *is* asleep," Scott answered. And then we both realized what was happening.

The eel was dreaming.

In our dreams, we humans experience our most isolated and mysterious existence: "All men," wrote Plutarch, "while they are awake, are in one common world; but each of them, when he is asleep, is in a world of his own." How much more inaccessible, then, are the dreams of animals?

Humans have always exalted dreams. Pindar of Thebes, the Greek lyric poet, suggested that the soul is more active while dreaming than while awake. He believed that during a dream, the awakened soul may see the future, "an award of joy or sorrow drawing near." So it's no wonder that humans were quick to reserve dreams for people alone; researchers for many years claimed dreams were a property of "higher" minds. But any pet owner who has heard her dog woof or seen his cat twitch during sleep knows that is not true. MIT researchers now know not only that rats dream, but what they dream about. Neurons in the brain fire in distinctive patterns while a rat in a maze performs particular tasks. The researchers repeatedly saw the exact same patterns reproduced while the rats slept—so clearly that they could tell what point in the maze the rat was dreaming about, and whether the animal was running or walking in the dream. The rats' dreams took place in an area of the brain known to be involved with memory, further supporting a notion that one function of dreams is to help an animal remember what it has learned.

A 1972 study mistakenly suggested that the platypus, a primitive, egg-laying mammal whose ancient lineage stretches back 80 million years, did not experience REM sleep, the kind of sleep during which humans dream. But those researchers were looking in the wrong place in the brain. In 1998, a new study showed that, in fact, the platypus experiences more REM sleep—some fourteen hours a day—than any other known mammal.

Much less work has been done on fish than on mammals. But it is known that fish sleep. Even nematodes and fruit flies sleep. A 2012 study showed that if fruit flies' sleep is interrupted repeatedly, they have trouble flying the next day—just as a person would have trouble concentrating after a sleepless night.

In a book I love so much that my husband reads it to me every Christmas, Wales's greatest poet, Dylan Thomas, takes his readers to Milk Wood, a small town by the "slow, black, crowblack, fishing-boat-bobbing sea." It is night, and the characters in the book are all asleep; the author offers his readers a chance to enter the most alluring and impossible of intimacies: "From where you are," he promises, "you can hear their dreams."

When a fish appears to you in a dream, according to Jungian interpretation, the animal represents insights bubbling up from the intimate, oceanic mystery of the unconscious. But on this morning, on an ordinary day in a public institution, while moms pushed their babies in prams and kids laughed and pointed and squealed around me, I had experienced not just an insight, but a revelation: I had seen the dreams of a fish, hunting and stunning its prey.

❧

We returned to Octavia, and Scott put the squid on the end of the long grabber so he could hold the food right in front of her face. She seized the squid—and the tongs. I ran up the steps to the tank, stubbing my toe, and plunged both arms into the water. She dropped the squid. She'd wanted the tongs—and now she wanted me, too. While holding fast to the side of the tank with hundreds of suckers, and still holding the tongs in dozens of others, Octavia grabbed my left arm with three of her arms and my right arm with yet another of hers, and began to pull—hard.

Her thorny red skin showed her excitement. Her suction was strong enough that I felt her drawing the blood to the surface of my skin. I would go home with hickeys that day. I tried to stroke her, but my hands were immobilized. She kept me at arm's length, but at least I could see her head. It was now the size of a cantaloupe, and each arm was at least three feet long. She had grown dramatically since my previous visit. The giant Pacific octopus is one of the world's most efficient carnivores in converting food to body mass. Hatching from an egg the size of a grain of rice weighing three-tenths of a gram, a baby giant Pacific octopus doubles its weight every eight days until it reaches about 44 pounds, then doubles its weight every four months until maturity.

Scott was pulling with all his considerable strength on the tongs to keep Octavia from pulling me into the tank. I submitted to the tug-of-war. I had no choice. Though fairly fit for a person of my size (five foot five, 125 pounds), age (fifty-three), and sex (female), I didn't have the upper-body strength to resist Octavia's hydrostatic muscles. An octopus's muscles have both radial and longitudinal fibers, thereby resembling our tongues more than our biceps, but they're strong enough to turn their arms to rigid rods—or shorten them in length by 50 to 70 percent. An octopus's arm muscles, by one calculation, are capable of resisting a pull one hundred times the octopus's own weight. In Octavia's case, that could be nearly 4,000 pounds.

Though octopuses are usually gentle, there are accounts of people who have drowned, or nearly so, as a result of the animals' attentions. English missionary William Wyatt Gill spent two decades in the South Seas, among octopuses much smaller than the giant Pacific; but even these species are strong enough to overwhelm a young, strong, fit man. He wrote that "no native of Polynesia doubts the fact" that octopuses are dangerous. Gill reports that one fellow, who was hunting octopus, would have smothered were it not for his son, who rescued him when he surfaced with an octopus blanketing his face.

Another account comes from waters off New Zealand from a D. H. Norrie, who was wading in sea channels searching for lobsters with Maori friends. Suddenly, one of his companions "began shrieking and trying to free himself from something that was holding him fast. We moved over to help him and found him to be struggling with a young octopus!" The animal was only 30 inches long—and yet without his friends, Norrie told the author Frank Lane, the man would never have escaped, and would have surely drowned.

Octavia was using only a tiny fraction of her great strength. Compared to what she *could* do, this was just a playful tug. I didn't feel I was under attack. I felt I was under investigation.

I could have been in her grip for only a minute, or it might have been five, but after what felt like a considerable time, suddenly she shrank from us. She let go of me and the tongs at the same moment.

"Wow!" I said as she retreated to her lair. "That was amazing!"

"I was pulling with all my strength!" Scott said. "I was afraid I would end up holding you by the ankles!"

What had happened between Octavia and me? What was she thinking? It was obvious she wasn't hungry, or she would have eaten the squid. She didn't seem fearful or angry—I can almost always feel that from a mammal or bird, though I wasn't sure I could pick it up from a mollusk. Yet Scott and I agreed that this encounter was utterly different from my first, playful encounters with Athena. "This may have been some sort of dominance display," Scott said. Perhaps she wanted the tongs and concluded, reasonably though incorrectly, that I was keeping them from her. Another thought occurred to me: When I stubbed my toe racing up to the tank, my chemistry changed, as the neurotransmitters associated with pain flooded into my system. Being able to recognize the neurotransmitters of pain would be a useful ability for an octopus; then it could tell whether prey was injured and therefore particularly easy to subdue. Earlier in the day, I had seen a fish's dreams—and now, perhaps, an octopus had tasted my pain.

In this watery realm, I was being drawn to possibilities I had never before imagined.

✻

Those who work with octopuses report seeing things that, according to the way we've learned the world normally works, should not be happening.

Such was the day Alexa Warburton found herself chasing a fist-size octopus as it ran across the floor.

Yes, *ran.* "You'd chase them under the tank, back and forth, like you were chasing a cat," she said. "It's so *weird!*"

Alexa was a pre-veterinary student at Middlebury College's newly created octopus lab in Vermont. It seemed to her that some of the octopuses were purposely, and sometimes elaborately, uncooperative. When a student would try to scoop an animal from its tank with a net and transfer it to a bucket to run a T-maze, for example, the octopus might hide, squeeze into a corner, or hold fast to some object and refuse to let go. Some would allow themselves to be captured, only to use the net as a trampoline. They'd leap off the mesh like acrobats and dive to the floor—and then run for it.

Alexa described the experience of working with these small invertebrates as "surreal." At the little lab, which was located in a former janitorial closet, she and the other students worked with two different species: the tiny Caribbean dwarf octopus, and the larger California two-spot, which can reach a mantle length of seven inches and have arms up to 23 inches long. "They were *so* strong," she said. "This animal is so small, it fits in my hand—and yet it's as strong as I am!"

The lab's 400-gallon tank had a weighted lid, and was divided into separate compartments for each animal. But the octopuses would escape. They would push out from under the lid and crawl out, and sometimes die; they would dig beneath the dividers, which the students had hammered in, to get in another octopus's compartment, and they'd eat each other. Or they'd mate, which was just as lethal for the students' experiments. After mating, females lay eggs, hole up, and refuse to run mazes, and then when their eggs hatch they die; the males die soon after mating.

Even more impressive than the octopuses' physical strength was the force of their will, the sheer strength of each individual personality. The students were supposed to refer to their animals by numbers in their research papers, but they ended up calling them by name: Jet Stream, Martha, Gertrude, Henry, Bob. Some were so friendly, Alexa said, "they would lift their arms out of the water like a dog jumps up to greet you"—or like a child who wants to be lifted up and hugged. One named Kermit liked Alexa to pet him, and seemed to snuggle into the caress "by raising his shoulders—even though he didn't have shoulders."

Others were irascible. One of the dwarf Caribbeans was such a problem the students called her the Bitch. "Catching her for the maze always took twenty minutes," Alexa said. This octopus would invariably grip onto something and not let go.

And then there was Wendy. Alexa used her as part of her thesis presentation. It was a formal event that was videotaped, for which Alexa wore a nice suit. As soon as the cameras started rolling, Wendy drenched the student with salt water. Then the octopus scurried to the bottom of the tank, hid in the sand, and refused to come out. Alexa is convinced the whole debacle occurred because the octopus realized in advance what was going to happen and resolved to prevent it.

"Wendy," she said, "just didn't feel like being caught in the net."

Data from Alexa's experiments showed the California two-spots were quick learners. But Alexa learned far more than a refereed journal could publish. "They're so curious," she told me. "They want to know about everything around them. An invertebrate! This supposedly simple, simple animal!"

"We don't understand them," she continued. "Try to make a maze that will show how this creature thinks. We don't even understand them enough to test them. Maybe mazes aren't the way to study them. Science can only say so much. I *know* they watched me. They followed me. But proving that intelligence is so difficult. There's nothing as peculiar as an octopus."

⁂

A week after Octavia nearly pulled me into her tank, I was back at the aquarium.

Sparked by an article I had written for *Orion* magazine, friends at the national environmental radio show *Living on Earth* wanted to record a segment with me on octopus intelligence. They hoped to interact with Octavia. I had no idea what to tell them to expect.

I came in early to visit with Scott, Wilson, and Bill. What kind of reception might Octavia give my radio friends? Bill, who had worked with five octopuses over his eight years at the aquarium, characterized her personality this way: "Aggressive and standoffish."

"This one," agreed Wilson, "isn't playful." Unlike all the others, he said, about half the time he tried to interact with her, she completely ignored him.

Octavia was different from the other octopuses Wilson had known in another important way: She camouflaged. Previous octopuses, who had all come as young pups, had lived behind the scenes in tanks or barrels that were completely barren—no hiding places, no rocks or sand or tank mates. And though these octopuses could turn color—growing red when excited, pale or white when calm, and showing shades of brown and white and mottling in between—they didn't camouflage to match their background. There was nothing much to match. Wilson noticed that even when they were transferred to the public display tank, they still didn't camouflage.

But Octavia did.

The ability of the octopuses and their kin to camouflage themselves is unmatched in both speed and diversity. Octopuses and their relatives put chameleons to shame. Most animals gifted with the ability to camouflage can assume only a tiny handful of fixed patterns. The cephalopods have a command of thirty to fifty different patterns per individual animal. They can change color, pattern, and texture in seven tenths of a second. On a Pacific coral reef, a researcher once counted an octopus changing 177 times in a single hour. At Woods Hole Oceanographic Institution, cephalopods put on laboratory checkerboards virtually disappear. They don't make checks, of course; but they can create a pattern of light and dark that makes them invisible on virtually any background, to virtually any eye.

Octopuses and their relatives have what Woods Hole researcher Roger Hanlon calls electric skin. For its color palette, the octopus uses three layers of three different types of cells near the skin's surface—all controlled in different ways. The deepest layer, containing the white leucophores, passively reflects background light. This process appears to involve no muscles or nerves. The middle layer contains the tiny iridophores, each 100 microns across. These also reflect light, including polarized light (which humans can't see, but a number of octopuses' predators, including birds, do). The iridophores create an array of glittering greens, blues, golds, and pinks. Some of these little organs seem to be passive, but other iridophores appear to be controlled by the nervous system. They are associated with the neurotransmitter acetylcholine, the first neurotransmitter to be identified in any animal. Acetylcholine helps with contraction of muscles; in humans, it is also important in memory, learning, and REM sleep. In octopuses, more of it "turns on" the greens and blues; less creates pinks and golds. The topmost layer of the octopus's skin contains chromatophores, tiny sacks of yellow, red, brown, and black pigment, each in an elastic container that can be opened or closed to reveal more or less color. Camouflaging the eye alone—with a variety of patterns including a bar, a bandit's mask, and a starburst pattern—can involve as many as 5 million chromatophores. Each chromatophore is regulated via an array of nerves and muscles, all under the octopus's voluntary control.

To blend with its surroundings, or to confuse predators or prey, an octopus can produce spots, stripes, and blotches of color anywhere on its body except its suckers and the lining of its funnel and mantle openings. It can create a light show on its skin. One of several moving patterns the animal can create is called "Passing Cloud" because it's like a dark cloud passing over the landscape—making the octopus look like it's moving when it's not. And of course the octopus can also voluntarily control its skin texture—raising and lowering fleshy projections called papillae—as well as change its overall shape and posture. The sand-dwelling mimic octopus, an Indonesian species, is particularly adept at this. One online video shows the animal altering its body position, color, and skin texture to morph into a flatfish, then several sea snakes, and finally a poisonous lionfish—all in a matter of seconds.

No researcher today suggests that all of this is purely instinctive. An octopus must choose the display it needs to produce for the occasion, then change accordingly, then monitor the results—and, if necessary, change again. Octavia's camouflage abilities were superior to those of her predecessors because, living longer in the ocean among wild predators and prey, she had *learned* them.

This is more evidence of the octopus's alien, invertebrate intelligence. But I feared my friends from the radio show might not get to see even a glimmer of Octavia's sparkling mind, and would witness only a baggy, boneless body balled up in its lair. "If she doesn't want to come," Wilson reminded me, "forget it."

So I was completely unprepared for what happened when Bill opened the top of her tank that afternoon. As host Steve Curwood, his producer, and his sound crew stood by, Wilson fished a capelin out of the small plastic bucket of fish he had positioned at the lip of Octavia's tank. Flushed with excitement, Octavia flowed immediately over to him—not just extending an arm or two, but rushing toward him with her whole body. Her head bobbed to the surface so she could look into our faces. She looked us both directly in the eyes and then accepted the capelin. As she conveyed it toward her mouth, three of her arms rose from the water, and she grabbed Wilson's free hand with some of her largest suckers. I plunged my hands and arms in and she grabbed me, too. One arm, two, then a third, attached to me. I could feel the suction of the suckers, but her arms did not pull me.

"Steve, meet Octavia." Bill invited Steve to let her touch him, too. "Roll up your sleeves; take off your watch," he instructed. "We always joke that they're very sticky fingered, so they could probably slip off a ring or a watch without you realizing it, but also, we don't want anything sharp on ourselves that would hurt them."

Steve obliged and extended his fingers. Octavia uncurled an arm to taste him.

"Oh!" cried Steve. "She's grabbing ahold, here—"

Wilson handed Octavia another capelin.

"Yup, feel the suckers!" said Steve. Bill explained that she could control each sucker individually. "Wow!" said Steve. "So she'd be amazing playing the piano—can you imagine?"

We were drowning in sensation: the feel of Octavia's suckers on our skin, the spectacle of her subtly changing color, the procession of the capelin as they were passed forward to her mouth, the unconstrained acrobatics of her many unjointed arms. Six of us were watching her, and three of us had arms in the tank, before anyone noticed what had happened: She had managed to steal the bucket of fish right out from under us. She was holding the bucket fast with some of her strongest, biggest suckers while using hundreds of other suckers to explore Wilson, Steve, and me.

Octavia wasn't interested in the fish. They were still in the bucket. She was holding it in such a way that the fish, in the bucket's bowl, faced away from her. She drew the webbing between her arms around the bucket almost like a hawk hides its captured prey with its wings. As she had been with the tongs she grabbed from Scott the week before, Octavia was more interested in the object that held the food than in even the food itself.

Apparently the six of us were not sufficiently interesting to occupy her vast capacity for attention. And unlike the guest who texts and checks his e-mail while eating and carrying on a conversation at a dinner party, Octavia did not seem distracted as she multitasked; she was able to focus on each of her many, simultaneous efforts. This stunned us all the more, because we had clearly been overcome by our single—and, one would think, simple—task: watching what the animal, who we were actually touching, was doing.

"So if an octopus is this smart," Steve asked Bill, "what other animals are out there that could be this smart—that we don't think of as being sentient and having personality and memories and all these things?"

"It's a very good question," Bill answered. "Who knows what else is actually out there in the ocean?"

⚜

For an invertebrate, the octopus brain is enormous. Octavia's was about the size of a walnut—the same size as that of an African gray parrot. Alex, an African gray trained by Dr. Irene Pepperberg, learned to use a hundred spoken English words meaningfully;

demonstrated an understanding of concepts of shape, size, and material; could do math; and asked questions. He could also purposely deceive his trainers—as well as apologize when he was found out.

Brain size, of course, isn't everything. After all, anything can be miniaturized, as computer technology plainly shows. Another measure scientists use to assess brain power is to count neurons, the mainstay of the brain's processing capabilities. By this measure, the octopus is again impressive. An octopus has 300 million neurons. A rat, 200 million. A frog, perhaps 16 million. A pond snail, a fellow mollusk, at most, 11,000.

A human, on the other hand, has 100 *billion* neurons in the brain. But our brain is not really comparable to that of an octopus. "Short of Martians showing up and offering themselves up to science," says neuroscientist Cliff Ragsdale of the University of Chicago, "cephalopods are the only example outside of vertebrates of how to build a complex, clever brain." Ragsdale is investigating the neural circuitry of the octopus brain, to see if it works at all like ours.

The human brain, for instance, is organized into four different lobes, each associated with different functions. An octopus brain, depending on the species and how you count them, has as many as 50 to 75 different lobes. And most of an octopus's neurons aren't even in the brain but are in the arms. These may be adaptations for the sort of extreme multitasking an octopus must undertake: to coordinate all those arms; to change color and shape; to learn, think, decide, and remember—while at the same time processing the flood of taste and touch information pouring in from every inch of skin, as well as making sense of the cacophony of visual images offered by the well-developed, almost humanlike eyes.

But like our eyes, our brain and the octopus brain arrived at their complexity by different routes. The common ancestor of humans and octopuses—a primitive, tube-shaped creature—lies so deeply embedded in the prehistoric past that neither brains nor eyes had yet evolved. Still, the octopus eye and our own are strikingly similar. Both have lens-based focusing, with transparent corneas, irises that regulate light, and retinas in the back of the eye to convert light to neural signals that can be processed in the brain. Yet there are also differences. The octopus eye, unlike our own, can detect polarized light. It has no blind spot. (Our optic nerve attaches to the back of the eye at the retina, creating the blind spot. The octopus's optic nerve circles around the outside of the retina.) Our eyes are binocular, directed forward for seeing what's ahead of us, our usual direction of travel. The octopus's wide-angle eyes are adapted to panoramic vision. And each eye can swivel independently, like a chameleon's. Our visual acuity can extend beyond the horizon; an octopus can see only about eight feet away.

There is another important difference as well. Human eyes have three visual pigments, allowing us to see color. Octopuses have only one—which would make these masters of camouflage, commanding a glittering rainbow of colors, technically *color-blind*.

How, then, does the octopus decide what colors to turn? New evidence suggests cephalopods might be able to see with their *skin*. Woods Hole and University of Washington researchers found the skin of the octopus's close relative, the cuttlefish *Sepia officinalis*, contains gene sequences usually expressed only in the retina of the eye.

Assessing the mind of a creature this alien demands that we be extraordinary flexible in our own thinking. Marine biologist James Wood suggests our hubris gets in our way. He likes to imagine the way someone like Octavia might attempt to measure our brainpower: "How many color patterns," he suggests an octopus might wonder, "can your severed arm produce in one second?" On the basis of that answer, Octavia could reasonably conclude we humans were stupid indeed—so dumb that she could steal a bucket of fish from us in full view. The thought was humbling. But so was a possible alternative. Roman natural historian Claudius Aelianus observed of the octopus in his writing at the turn of the third century that "mischief and craft are plainly seen to be the characteristics of this creature." Perhaps Octavia had recognized our intelligence, and enjoyed her bucket all the more for having outwitted us.

⁂

Each time I visited her thereafter that fall and winter, Octavia rose to the top of the tank and flowed over to meet me, eager to taste me with her suckers and look me in the face. Sometimes I brought a friend. Not only was I eager to share this experience, I also wanted to see how Octavia reacted to other people. She met my friend Joel Glick, a nonsmoker, who had studied mountain gorillas in Rwanda and would soon be leaving to study a colony of imported macaque monkeys in Puerto Rico. Octavia embraced Joel wholeheartedly.

One December day I brought a high school senior, Kelly Rittenhouse, an aspiring writer. We had never met before, but she had read some of my books and asked to job-shadow me for a school project. On our drive to Boston, I told Kelly I was a little worried about my hair. I had gotten a perm earlier in the week. I feared Octavia might taste any chemicals that might have seeped into my skin and blood and not want to interact with me.

But Octavia came to me right away and quickly immobilized both my arms with her suckers. Scott had to keep pulling them off my skin. After a few minutes, when she seemed to calm down, we invited Kelly to touch her. Octavia began to tentatively taste Kelly with the suckers of one arm. And then—

Explosion! The rolled-up sleeves of my shirt and the top of my pants were suddenly wet. I looked up at Kelly to see water dripping

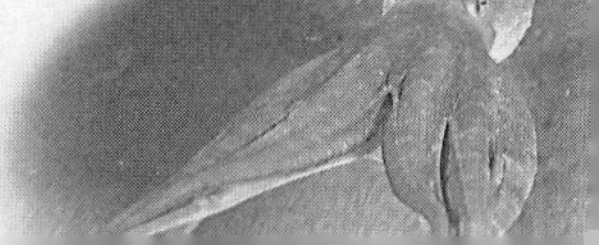

off her dark brown bangs and glasses and down her nose. Octavia had blasted her squarely in the face.

Kelly was doused. Her sweater was drenched. And even though her hosing made the three-block walk back to my car freezing cold, Kelly couldn't stop smiling. Later she e-mailed to tell me that her day had been "crazy awesome."

❧

Why had Octavia hosed Kelly? It's well known that octopuses use their funnels to repel what they don't like. They'll shoot jets of water at food detritus in front of their dens. They also blast water to express dissatisfaction. One common octopus who was part of a learning experiment in the 1950s so despised the lever that he was supposed to pull to get food, that he soaked his experimenter each time he was presented with it. (He eventually pulled the hated lever out of the tank wall.) But they also squirt for another reason: to play.

My first indication of this came after I wrote about the New England Aquarium volunteer whom Truman constantly squirted. She read the article and contacted me to say that she had enjoyed the article, but wanted me to know that Truman had *not* disliked her. The two of them had been friends. She so treasured the memory of her time with Truman that it was important to her that I understood.

Maybe, I thought, the octopus had squirted her in the same spirit of little boys who pull girls' pigtails, or the way kids might splash each other in the pool. Maybe the octopus was just teasing.

And then I met Jennifer Mather and Roland Anderson.

Jennifer, a psychologist at the University of Lethbridge in Canada, is one of the world's leading researchers on octopus intelligence; Roland, a biologist at the Seattle Aquarium, is another. Together and separately, they have scientifically investigated the octopus mind, exploring problem-solving and personality—even developing a personality test, using nineteen different, distinctive behaviors, to rank octopuses from shy to bold.

Roland made one of the team's most important discoveries one day while he was conducting an experiment on octopus preferences. Eight octopuses in separate three-by-two-by-two-foot tanks at the Seattle Aquarium's holding area were presented with empty Extra Strength Tylenol pill bottles. (Roland found octopuses can open the childproof caps, an achievement that eludes many PhDs.) "Some of the bottles had been painted white and others black; on some, the epoxy paint had been sprinkled with sand, to see if they preferred dark or light, smooth or rough," Roland, a dapper, slender man with a trim silver moustache, told me. "The bottles were weighted with rocks so they barely floated. We'd feed them one day and test them the next. How long does the animal hold on to different colors and textures? I watched what they did."

Some grabbed the bottle, explored it, and cast it off. Others grasped the bottle with one or two suckers and held it at arm's length, as if examining the object with suspicion. But two did something very different. They squirted it with their jets—but in a way Roland had not seen before. "This was not a strong, forceful jet," like you would use to squirt an irritating researcher, Roland explained, "but one carefully modulated so that the pill bottle was caused to circle around the tank over and over. She repeated the action sixteen times!" By the eighteenth time, he was already on the phone with Jennifer with the news: "She's bouncing the ball!"

A second octopus in the study later used her jet in a similar way, only sending the bottle back and forth across the water's surface instead of around the tank. They were both using their funnels—organs originally evolved for respiration and locomotion—to play.

The study was published in the *Journal of Comparative Psychol-*

ogy. "It fit all the criteria for play behavior," Roland told me. "Only intelligent animals play," he stressed. "Birds like crows and parrots; primates like monkeys and chimps; dogs and humans."

Perhaps this was what Octavia was doing with Kelly; perhaps this was what Truman was doing with the young volunteer he always squirted. Jennifer once saw a Pacific day octopus squirt at a butterfly flying over it in Hawaii; the butterfly, alarmed, had hurried away. Perhaps the octopus was annoyed at the shadow the butterfly had cast; or perhaps, in the manner of children who like to run at pigeons strutting in a public square to watch them scatter, the octopus had done it just for fun.

❧

I met Jennifer and Roland at the Octopus Symposium and Workshop at the Seattle Aquarium, to which both Bill and I had traveled to attend. The symposium—so successful that by its end, the organizers were already planning a second—was a revelation. In a large meeting room of an upper floor of the Seattle Aquarium, sixty-five octopus lovers, from internationally respected researchers to home hobbyists, gathered from at least five countries for ten presentations from experts on their favorite animal. "How many of you keep octopuses?" Jennifer asked the crowd during her keynote speech, the first after Roland's introduction. About fifty hands went up. "And do they have personalities?" Like a unanimous vote at a town meeting came the emphatic answer: "Yes!"

The first night in Seattle, Bill and I had dinner with Jennifer, a silver-haired éminence grise with rosy cheeks, a professor's thick glasses, and a quick smile. Joining us were other experts: David Scheel, an Alaska Pacific University professor and researcher; Gary Galbreath, an evolutionary biologist from Northwestern University; and David's student, Rebecca Toussaint. Rebecca would announce a stunning discovery the next day: Genetic testing shows that at least two distinct species of giant Pacific octopus exist in Alaskan waters, and perhaps elsewhere as well. The giant Pacific octopus, as Jennifer would point out, may be the Octopus Archetype, the ur-octopus, the ultra-octopus, the octopus every kid who's ever visited a public aquarium knows. Yet there are actually two distinct species, which dramatically underscores just how little science knows about these charismatic but mysterious animals.

Octopus experts like to discuss matter-of-factly some horrible things you find in the ocean. Jennifer told us about a transparent, stinging hydroid she'd encountered in Bonaire: "You can't see or predict where it's going to be," she said. Rebecca remembered the time a fire coral had brushed her elbow on a dive. "At first it didn't hurt," she said. "Then I got out, and I thought I would die!"

They also told us about Paul, the octopus from Sea Life Oberhausen, in Germany, who correctly predicted the outcomes of the 2010 FIFA World Cup soccer matches seven times in a row. Before a match, Paul would be offered two boxes, each containing a mussel. Each was adorned with a different flag representing the two nations whose teams would face one another in the upcoming game. How did Paul make his selection? And how did he do so with such success? We considered the possibilities—including that the octopus was drawn to the aesthetic qualities of one flag over the other, and that he really did know which team would win.

That night Jennifer and David also discussed a possible field expedition to investigate food preferences and personality in Pacific day octopus. Maybe, they said, I could go along.

⚜

After the Octopus Symposium, when I saw Octavia again, she held on to me, gently but firmly, for an hour and fifteen minutes. I stroked her head, her arms, her webbing, absorbed in her presence. She seemed equally attentive to me. Clearly, each of us wanted the other's company, just as human friends are excited to reunite with each other. With each touch and each taste, we seemed to reiterate, almost like a mantra: "It's you! It's you! It's you!" Finally Bill and Scott asked me to stop so they could close the lid and we could go to lunch. Though my hands had frozen, I hated to leave, especially since I was soon due to go on a book tour and wouldn't get to see Octavia again for two months.

And though I travel extensively and often, this time I found being on the road exceptionally difficult. This time, my usual home-sickness was compounded by being away from the octopus.

When I returned, I e-mailed Bill to see when I could visit. Bill wrote me back warmly, but with alarming news:

"Octavia is being temperamental because she's getting old, so hopefully she will come out to say hi. . . ."

Getting old? I felt sick. Could her life end so soon and so suddenly, like Athena's?

Jennifer had warned me: "If an octopus lives long enough, it becomes senescent. I'm reluctant to use the word *dementia*—it's so human-specific and associated with mental illness, and it's not normal or natural or inevitable for every person who lives long enough. But senescence happens to every octopus who does."

Alexa had witnessed this decline in the octopuses at Middlebury when they aged. "They swim loop-the-loop in the tank, they look all googly-eyed," she said. "They won't look you in the eye or attack prey." One senile octopus at the lab crawled out of the tank, squeezed into a crack in the wall, dried out, and died.

When senescence strikes the larger species, like the giant Pacific, the results can be even more dramatic. One day when James Cosgrove was working as a display diver for the Pacific Undersea

Gardens in Victoria, British Columbia, he was attacked by a huge male—in full view of the delighted public. The facility is a floating aquarium where guests descend 11 feet below sea level and divers bring interesting animals up to the windows to show the visitors. The diver checked a cave-like entrance near the entry ladder and discovered what he thought were two octopuses inside—but as the arms slithered past his mask, showing enormous suckers, he realized he had instead found one monster octopus, who then seized him. "All I could do was keep both hands on my regulator [the mouth-piece that delivers the diver's air supply] while the octopus dragged me around like a sack of potatoes," he wrote in *Super Suckers*. "At one point I could see that the octopus could reach from the display windows to the outside screen, which was a distance of 22 feet." A few weeks later the octopus died. He had weighed 156 pounds. Cosgrove concluded the octopus was out of his mind.

Neither Scott nor Bill could remember a senescent octopus at the aquarium becoming aggressive. They usually just turn unresponsive and vacant, which, Bill told me when he met me in the aquarium lobby the next day, was happening now to Octavia. "Three weeks ago her behavior changed," he said. "Usually, you know, she's in the top corner of her tank. Now she sits on the bottom or in the window by the brighter light. She eats, but she takes the food and runs it back to her corner. Sometimes she won't come over at all. She just sends out an arm. In the mornings, she seems really white. She was always an exceptionally red octopus. But she's now faded. She's pale."

This surely must have pained Bill. "She turned out to be a really friendly, interactive octopus," he said, as if mourning her passing already. Shortly before her senescence kicked in, some federal agents had come by to deliver an illegally imported arowana—a long, thick, silvery ribbon of a fish kept in aquariums throughout Asia for good luck—they had confiscated. Scott had invited the agents to interact with Octavia as a thank-you. Octavia had seemed particularly interested in one agent, and her arms were all over him. Then she started to pull. "And I saw this look on his face," Scott said, "on the threshold of panic." Then it occurred to Scott: "Most of these guys have sidearms." Octavia might have been reaching for his gun, curious about this new object. "Now," he said, "*that's* enrichment!"

"Are you sure your safety's on?" Scott asked the officer. He quickly disentangled the agent from Octavia's grip. "That's not a story we want to be told," he said: "'Agent Shot in Foot by Octopus.'"

Not long after that, Octavia seemed to lose her zest for interaction. As much as I longed to see her, I dreaded seeing her decline. I had, of course, seen humans I loved in similar straits: A friend, a former trapper turned naturalist, had a stroke and babbled incoherently, not realizing that nobody could understand him as he held up his end of the conversation with great animation. Oddly, at one point when my husband and I visited him in the hospital, he suddenly spoke one sentence in clear English and said, "The deer—a buck—I dropped him on the run." My friend Liz's mother, Lorna, a ballerina turned anthropologist, lived to nearly 104; two years after Harvard published her first book, at age 102, she began to forget people's names. She forgot mine shortly after she turned 103, but she clearly remembered I was important to her and greeted me with genuine warmth. I had seen this in our first border collie, when she was sixteen. She'd wake my husband and me in the night, crying and frightened, as if she couldn't remember where she was or who we were. I'd lie on the floor with her and stroke her and kiss her till the light came back into her intense brown eyes, as if her soul had returned from a journey.

In all these cases, a piece of these individuals' minds had gone missing. Had their selves gone missing with it? Who *were* they now?

And what does an aging octopus like Octavia experience in this phase in the life of her multifaceted mind?

"I hope she'll lay eggs," Bill said to me as we headed to Octavia's tank. "That's a sign she could live six more months." Even in a diminished mental state, we wanted Octavia to stay with us, just as I had wished for my friends and our dog, even after pieces of their souls seemed to be falling away. "And after we see Octavia," Bill promised, to cheer us both up, "I have a surprise."

Bill opened the tank and offered Octavia a shrimp on the long tongs. She sent forth an arm, suckers up—and then another arm came over, followed by the rest of her body. I could see she was paler than usual. I reached out to some of her larger suckers, and she attached them to me, but weakly. Next Bill gave her a capelin. The sea star leaned in, sensing the food. I gave Octavia both my arms and she tasted me with four of hers, while conveying the capelin toward her mouth. Bill pointed out a three-quarter-inch crescent of white, ragged flesh between the webbing of her second and third arms. It wasn't just pale; it looked necrotic. Rather than moist, healthy octopus skin that belonged in the water, it looked like a piece of sodden Kleenex that had somehow ended up here by mistake and was coming apart. It looked like she was disintegrating, leaving this world piece by piece.

I looked up and saw Wilson coming down the wet hallway of the Cold Marine Gallery. I was so glad, as I hadn't seen him since December—five months ago—and that period had been a very trying time for both Wilson and Scott.

In December, Scott had lost one of his favorite animals, an arowana he had raised from a baby and known for years, and one of his electric eels had been at fault. Behind the scenes, moved to a temporary tank so his regular tank could be cleaned, the big fish had leapt into an adjacent tank and electrocuted both Scott's beloved arowana and another valuable animal, an Australian lungfish. That same month, Wilson had a major operation on his back.

Meanwhile, as Wilson recovered from his operation, his wife, an accomplished social worker with a wry sense of humor, had been stricken by a neurological disease that was eroding both her muscles and her mind, a disorder that doctors could neither explain nor stop.

Wilson hadn't been to the aquarium but twice since December. On this May day, he made a special trip in from his home in Lexington, Massachusetts, to see me. He gave me a big smile and a hug.

I thought Wilson's presence was Bill's surprise for me. But it was not.

"So," Wilson said to me, "have you seen the new baby octopus?"

CHAPTER THREE

Kali

The Fellowship of the Fishes

Octopuses are famous for showing up in places that surprise you. One giant Pacific octopus took up temporary residence in a pair of overalls in a shipwreck (and scared a diver half to death when the overalls rose up before him, writhing). Octopuses have turned up inside large conch shells and in scientists' tiny oceanic measuring instruments. Red octopuses particularly like to den in stubby brown beer bottles.

But I never expected to find Bill's new octopus in a pickle barrel in the sump.

On my way to visit Octavia, I had walked right past the sump, normally filled only with recirculating seawater, without noticing the barrel. The 55-gallon container's screw top was fitted with a lid of fine mesh. Its sides had been drilled with hundreds of three-eighth-inch-diameter holes, through which the water of the sump could freely flow.

This is the only container in the aquarium Bill deemed sufficiently octopus-proof for a giant Pacific this small; her head and mantle combined are about the size of a small grapefruit.

Looking into the water of the sump, I can see the dark tips of the new octopus's arms, fine as dental instruments, questing out from the barrel's holes. She can extrude her arms almost like toothpaste.

Already some six inches of arms are poking through three holes. That's why the holes are three eighths of an inch. "Three quarters of an inch," says Wilson, who had drilled them, "and she'd be out."

Bill determined her sex only two days ago. You can tell by looking at the tip of the third right arm. If the arm has suckers all the way to the tip, you have a female. If not, the appendage is referred to as the hectocotylized arm, and the animal is male. The reason it takes a while to tell is that octopuses won't always let you examine this arm, especially the males. They tend to keep the tip—the ligula—balled up and protected, and for good reason: This is the specialized organ for placing the spermatophore inside the female. (But he doesn't put it between her "legs," or arms, because that's where her beak is. He puts it in her mantle opening—or, as Aristotle explained it, he "has a sort of penis on one of his tentacles . . . which it admits into the nostril of a female.")

At first, Bill admitted, he was a little disappointed about the new octopus's sex. He was hoping for a boy. "Females can be feisty," he explains. "Males are more easygoing." They're also, he says, easier to name: "Frank, Stewy, Steve—any name for a male octopus is funny. Naming females is more challenging." He came up with the name Guinevere for his first octopus because he had been watching the movie *King Arthur*.

This little female, though, has already won Bill over. She was wild a week ago, and yet, by the time he unscrews the top and lifts the lid, she's already at the surface, looking at the three of us curiously with limpid, slit-pupil eyes.

"What a darling little thing!" I cry.

"She is beautiful," Wilson agrees.

"We like her," Bill says, his smile crinkling the edge of his eyes.

Compared with Octavia and Athena, this octopus is an exquisite miniature. She is half the size Octavia was when she first arrived;

though it is impossible to pinpoint the age of an octopus (growth rates depend on many variables, including water temperature), Bill estimates she may be younger than nine months. Her arms are less than a foot and a half long. She is a size, perhaps, I can finally get my mind around.

At first, she is a deep, rich chocolate color, except for a light spot on her head. As she looks at us, she turns a lighter brown, mottled with beige. Light stripes now curve down from her eyes toward where her nose would be, if she had a nose, like the "tear streaks" on a cheetah.

The reasons for an octopus's color change are myriad. Of course, an octopus might change to match or blend with its surroundings and become invisible; it may also change to look like something other than an octopus (presumably something less tasty or more threatening). But other changes certainly reflect mood. Nobody has figured out what all the color changes mean. A few are known: A giant Pacific octopus who turns red is generally excited; a white one is relaxed. An octopus presented with a difficult puzzle for the first time often undergoes several rapid changes in color, like a person who frowns, bites his lip, and furrows his brow when trying to solve a problem. A nervous octopus takes special care to disguise its head and especially its eyes, and can create a variety of spots, bars, and squiggles to confuse a predator. The small, deadly poisonous, blue ring octopus of Australia flashes dozens of electric-blue namesake rings all over its body when threatened. Another disguise is known as the eyebar display, in which an octopus makes a thick, dark line extend at the outer edge of the eye from either end of each slit pupil, masking the roundness that is typical of an eye. In Jennifer and Roland's studies showing that octopuses recognize individual humans, they found that after only a few trials, when the octopuses saw one of the staff members who always touched them with a bristly stick, they would make the eyebar as soon as they saw that person approach. When approached by people who always fed them, they did not.

But the white spot on the new octopus's head remains constant, even as our new octopus turns a deeper and more uniform brown again. Bill confirms that he has never seen her without this mark. Finally! A feature that stays the same on an octopus!

The spot reminds Bill of the bindi, the dot with which a lady decorates her forehead in India. So he named her Kali, after the dark-skinned, many-armed Hindu goddess of creative destruction. Like octopuses, the Hindu gods and goddesses are always changing form. When Kali takes the form of Prakriti, or Mother Nature, she dances upon the field of Consciousness (pictured as the supine body of her husband, Lord Shiva) with wild abandon. In other depictions, she wears a garland of skulls. Kali is a great name for this outgoing youngster, with her astonishing octopus powers and potentially destructive bent.

Wilson and I each offer her a finger, then a hand. She grips us gently with suckers of her two front arms.

"She's going to be friendly," says Wilson.

"Yes," says Bill, "she's going to be a good one."

⚶

Kali arrived just in time. In my quest to get to know an octopus better, I had been looking into acquiring one of my own.

Haunting cephalopod forums like TONMO.com (The Octopus News Magazine Online) and surfing the web, I was enchanted with videos posted by doting octopus owners. Some pet octopuses were wonderfully interactive. One person posted a video of a California two-spot *hopping*, on rear arms, back and forth across the sandy bottom of the tank, wildly waving its front arms at the front of the tank—looking for all the world like an eager student desperate for the teacher to call on him. The owner wrote that the octopus often did this to entice him to play. Later I read of a pet octopus who developed another way to signal that she wanted the owner's attention. If the person was out of the room, the octopus would pull off the magnet on the inside of the tank which, with another magnet on the outside of the tank, held a glass-cleaning tool in place. The outside magnet would then crash loudly to the ground, summoning the human much as one might call a butler by ringing a servant's bell.

Ceph keeper Nancy King discovered that her two-spot, Ollie, didn't always see where the live crabs she dropped in to feed her had landed. So she took to helping her, using her index finger on the outside of the aquarium to show her where the prey was hiding. Ollie soon figured out the meaning of the pointing finger. (This is a very specialized skill. Dogs—but not their direct ancestors, wolves—are among the tiny handful of species other than humans who can do this.) "In this way," she charmingly wrote, "Ollie and Nancy hunted crabs together."

Many home aquarists report that their octopuses appear to enjoy watching television with them. They particularly like sports and cartoons, with lots of movement and color. In their authoritative *Cephalopods: Octopuses and Cuttlefishes for the Home Aquarium*, King and her coauthor, Colin Dunlop, even suggest placing the tank in the same room as the TV, so owner and octopus can enjoy programs together.

But my husband was not enthused at the thought of an octopus in our house. In our nearly thirty years of marriage, he has successfully managed (so far) to fend off my getting snakes, iguanas, and tarantulas—as well as prevent me from keeping a red-tailed hawk for a falconry apprenticeship. He failed, however, to deflect a procession of other people's unwanted parrots from moving in with us, and once bought me a baby cockatiel, whom we both adored. We also adopted our landlord's cat, rescued two border collies, and raised baby chicks—in my home office, where they perched on my head and slept in my sweater. We even brought home a sick runt baby pig (who lived for fourteen years and grew to 750 pounds). My husband has loved them all, but his patience is often tested when I disappear into some jungle for weeks or months to research a book, and he's left with animals who invariably choose that moment to run away, try to kill each other, destroy their pens, roll in something, or throw up on the bed. Now an octopus?

When I brought up the subject, he replied, "Tell me this is a bad dream."

Expense aside—and it would cost thousands of dollars for the setup, food, and the octopus itself—there were logistical issues. Even for a small species like the Caribbean reef octopus, I would want a tank capable of holding 100 gallons of water. This would weigh at least 1,000 pounds—as much as a moose. And like a moose, its weight might collapse the floor of our 150-year-old farmhouse. Too, old houses like ours suffer from an insufficiency of electrical outlets, and a good saltwater aquarium tank needs several to run its complex life-support system: three kinds of filters, an aerator, and a heater to keep the water at the temperature that small tropical octopuses need, typically about 77° to 82°F.

In our neck of the woods, electricity itself is sometimes in short supply. We have frequent power outages, which last from minutes to days (in December 2008, after an ice storm, we had no electricity for a week), and even a relatively short period without filtration and heat can doom a tank and its occupants—especially if your octopus inks in alarm, which can poison the water, including the octopus itself.

Then there was the problem of proper water and food for an octopus. Natural seawater has more than seventy elements dissolved in it. The chemistry of the water must be exactly right for an octopus.

Any trace of copper, for instance, will kill it. And while an adult octopus will eat dead, frozen food, a very young one—which is what I would want, since the smaller species' life spans are even shorter than a giant Pacific's—needs live food. Since the nearest ocean is a two-and-a-half-hour drive from our house, I would need to raise the baby octopus's prey, amphipods and mysid shrimp, which would require their own separate aquarium setup.

Finally, if I had to travel (and I already had a research trip to Namibia scheduled that summer), my husband would end up with the perilous responsibility for the delicate octopus. In fact, the moment I left for Namibia, his work schedule would be subsumed by our border collie's struggle, after an operation on her tail, to defeat the Cone of Shame and chew out her stitches.

In the end, I decided that, as great as a personal home octopus might be, it would be too risky for both the octopus and my marriage. Besides, despite the long drive, I loved going to the aquarium. There, too, I had the benefit of being surrounded by experts—people whose observations would enrich and inform my own, people whom I now increasingly missed between visits. My plan, once I returned from Namibia, was to make more frequent trips to Boston to regularly observe Kali's growth and development. Wilson generously agreed to coordinate his schedule with mine. The week following my return from Africa, we inaugurated what we came to call our Wonderful Wednesdays, and dedicated this day each week to octopus observation. This provided an education both broader and deeper than I could have imagined, and sealed my connection not only with Kali but with the people who grew to love her just as much as I did—people who would become increasingly important in my life.

꙰

The next time I visit Kali, a small gaggle of staff and volunteers is already hanging around the sump as if gathered round the office coffeemaker. Except instead of sipping a hot beverage, they are dangling their hands casually in the freezing salt water, in order to hold hands with an octopus.

It's hard to imagine that Kali does not extrude her arms from the holes hoping for exactly this. In just two weeks, she has grown bigger, stronger, and more curious.

"She's bored," says Wilson as he unscrews the top of the barrel. She is already waiting for us at the top. "Correction," says Wilson, as she reaches up to taste his arm, "she *was* bored—she's not now!"

We offer Kali our hands and arms, and she latches on with eager suckers. You can almost feel her interest in the strong grip of her suction, as if she is eagerly reading us by using an octopus Braille system. And she wants to see as well as taste us. As her arms snake up over ours into the air, she lifts her head and eyes out of the water to look at us.

The slits of her pupils always remain horizontal, no matter what position she is in, cued by balance receptors called statocysts. These saclike structures are lined with sensory hairs and equipped with small, mineralized balls that shift inside the statocyst in response to motion and gravity. But the always-horizontal pupil can change dramatically in thickness. Under the bright light, you would think her pupils would be small, but now they are opened wide, like a person's when excited or in love.

Wilson hands her a fish—but she passes it *away* from her mouth. I find this astonishing in a rapidly growing young animal. Apparently her appetite for food is exceeded by her appetite for interaction. Kali wants to climb up our arms. Her glistening, muscular arm tips curl up over my forearm and my elbow and touch the cotton fabric of my shirtsleeve. Gently we pry away her suckers and urge her back into the water, but she grips us anew.

After a few minutes, Wilson breaks off the interaction. He doesn't want to overstimulate her. "She's still a baby," he says. "Let her rest."

Bill, who has been tending the feather duster worms (which take their name from the beautiful clusters of branched tentacles on their heads), tells us Kali recently entertained international visitors. The aquarium had hosted some staff members from the Beijing Aquarium. They were astonished to get a chance to touch an octopus, and even more astonished to find Kali so friendly. "They believed octopuses were very dangerous," said Bill.

Sea creatures in general evoke an irrational fear in many people, he has noticed. It's true that many of the creatures Bill cares for are venomous or have sharp teeth or poisonous spines. Yet all of the many scars on his long arms, he tells us, are from tubes, glass, and tools. "A screwdriver is more likely to draw blood from me than any of my animals," he says, laughing. "Yes, octopuses can bite. Yes, they can do damage. But people have a big fear of them, way out of proportion."

It's only relatively recently in the New England Aquarium's forty-year history that anyone dared interact with the octopus. Wilson tells me, "Fifteen years ago, nobody would go near the octopus."

Boston's aquarium was one of the first in the nation to offer naturalistic settings for its animals. It was a visionary change—one that not only made its exhibits more educational for the public, but also made them far more interesting for its animal inmates. With the exception of seals and sea lions (and of course the green sea turtle, Myrtle, who wouldn't stand for a brush-off), the policy of mirroring nature seemed to preclude much human interaction with the fish, reptiles, and invertebrates.

At lunch, Wilson and Scott tell me about the transformation that took place—one that was part of a quiet revolution throughout

zoos and aquariums and profoundly changed the relationships between people and the exotic animals in their care.

"It began with Marion," Wilson remembers. "Marion was mag-
nificent."

"Do you mean Marion anaconda or Marion Fish?" asks Scott.

Marion Fish—that really was her last name—had come first.
After retiring following twenty-six years working as a surgical
trauma nurse, in 1998 she started volunteering on Wednesdays and
got to know every animal she cared for personally. She named each
fish. And with remarkable accuracy, she could read their moods.

"She and I were sitting here with the octopus one day," Wilson
remembers, "and Marion said, 'You know, that octopus needs something to do.'" The idea of "enrichment"—of providing physical and
mental stimulation for zoo animals—was relatively new at the time,
even for chimps and tigers. It was unknown for fish and invertebrates. Direct contact with keepers was not part of the aquarium's
plan. "Back then, the others were afraid of touching the octopus,
afraid that touching the octopus would hurt it," Wilson tells me.
"But we said to hell with it. The octopus is bored! Then we started
playing with it." Soon Marion and Wilson were regularly opening
the tank to stroke the octopus and let the octopus suck on their
arms. It was clear the animal enjoyed the interaction, and maybe
even looked forward to the next encounter. "Then we gave it things
to play with—whatever was at hand. Tubes, things like that. That
was the start of everything," Wilson says. "Then I built the locking
cubes."

Marion Fish left the aquarium in 2003, after a heart attack, and
Scott and Wilson lost track of her. But in 2007, another Marion
appeared at the aquarium—a young woman whose influence was
equally profound. Marion Britt further demonstrated the positive
power of interesting, gentle, loving interaction between keepers and

the animals in their care. And she did it by directly handling the most fearsome animals in the aquarium—the 13-foot-long, 300-pound anacondas.

"Before Marion," says Wilson, "nobody would go into the tank with the anacondas." That sounded pretty reasonable to me. South America's top predators, anacondas readily hunt and kill adult deer, as well as 130-pound capybaras, and have been known to eat jaguars. I happen to have met one of the best-known biologists studying anacondas, Jesus Rivas, who has documented two predatory attacks by these powerful constricting snakes on his assistants in the field. Humans "are well within the predator-to-prey ratio" of anacondas, who can grow to 30 feet, he said. The only reason anacondas don't attack humans more often is that, other than Rivas and his field team, people don't venture where they know anacondas are found.

But Marion did. When she started at the aquarium as a twenty-four-year-old intern in Scott's gallery in 2007, there were three anacondas—whom nobody could safely touch. "We had to restrain the snakes whenever we handled them," Scott tells me. "We grabbed them behind the head. They hated it." By the time Marion stopped working at the aquarium, the two larger anacondas, Kathleen and Ashley, would slither up to her and curl up with their heads in her lap.

And now, thanks to Marion, no more are snakes traumatized by head restraint whenever they need to be moved from their tank for their yearly veterinary checkup, or to treat an illness, or when the tank needs to be drained. The staff no longer dreads interacting with them.

Clearly, the snakes are happier and healthier for it. The proof: Both females (the third, smaller snake, named Orange, turned out to be male) gave birth—the first time any anaconda in a Boston zoo or aquarium had done so. The soft eggs hatch inside the mother's body, and Marion was actually in the exhibit in her wet suit while Kathleen's seventeen babies were born. Since Marion handled all the baby snakes from both mothers since infancy, the offspring now on display, named Marion and Wilson (both female), don't need head restraint, either. With a little urging, they voluntarily submit to handling. The rest of the staff has also learned to recognize when the snakes are not in the mood to be handled, and back off at these times to try another day.

Marion had to leave the aquarium in February 2011 to have surgery, and because of surgical complications, hadn't returned. But the impact of her work has lasted. The sight of a slender young woman sitting in the anaconda exhibit with a 13-foot-long, predatory reptile snuggling in her lap, the tip of a tail coiled lovingly around one leg, provided dramatic evidence of what Scott and Wilson already knew: "Just about every animal," Scott says—not just mammals and birds—"can learn, recognize individuals, and respond to empathy." Once you find the right way to work with an animal, be it an octopus or an anaconda, together, you can accomplish what even Saint Francis might have considered a miracle.

Like Scott's latest project: training the Suriname toads.

Not only are these animals amphibians—with much less brain to work with than the anacondas—but they are also *blind*. Their blindness has sculpted their unique appearance: At the head of the toad's six-inch, flattened brown body are two nostrils, each set at the end of a long, narrow tube. The front limbs have star-shaped tactile organs on the fingertips to help the animals detect food.

Male Suriname toads call to their mates by clicking underwater, and the couple swim together in a series of circular loops while the female lays her eggs on the male's belly; the male fertilizes them and then rolls the eggs into pouches on the female's back. There the female's skin actually encloses the fertilized eggs to protect them.

When the female sheds her skin, the babies burst out from her back, pointy heads first. They are born not as tadpoles, but as perfect little toadlets.

Alas, the public seldom gets to see these exotic toads because the animals hide in the vegetation in their pretty, naturalistic exhibit. As he did with the electric eels, Scott is trying to figure out a way to induce the toads to show themselves.

How? "You need to get within the mind of the toad," he says. "We're engaged in toad psy-ops." How does a blind toad decide what is a safe, good place to stay—and how does he find it? "You get to learn very fast," Scott says. "You learn to project empathy. Remember the movie *E.T.*? It's kind of like that. You reach out with an invisible hand and read the organism. You have to meet them halfway. You have to be willing to listen."

Many of us respond without thinking to the angle of a horse's ears, or the position of a dog's tail, or the expression in a cat's eyes. Aquarists learn the silent language of fishes. Once, walking into a hallway behind the scenes where some cichlids had just been moved from one tank to another, Scott had announced to me with concern, "I smell fish stress." The scent is subtle—I cannot smell it at all—but the low-tide odor Scott detects, he explained at the time, is that of heat-shock proteins. These are intracellular proteins that were first discovered to be released, in both plants and animals, in response to heat, and are now known to be associated with other stresses as well. The scent makes Scott feel sick to his stomach—not because the smell is nauseating, but because the thought that fish in his care are stressed fills him with the same urgency and dread he used to feel when his newborn sons would cry.

Scott reads other fish cues just as fluently. When we visited the cichlids in their new home, he compared those who had just been moved to those who had been living there for weeks or months. The stripes on the new immigrants were paler. "And look at this one," he said, pointing to a fish who was already at home in the tank. "See the sparkle in the eye? Now look at this other one. You don't see the sparkle." Scott can read the faces of fishes as easily as you or I read a person's.

"The problem with reading octopuses," I say as we walk back to the aquarium, "is that they are *too* expressive"—much more than any species I'd ever known. "We have our poetry and dance and music and literature. But even with our voices and costumes and paint-brushes and clay and technologies, can we ever come close to ex-pressing what an octopus can say with its skin alone?"

"You're right," says Scott. "Imagine the road rage if cephalopods could drive on the Southeast Expressway!"

⁂

That afternoon, when Wilson opens her barrel, Kali bobs to the surface. Her eyes swivel, seeking our faces. We give her our arms and she embraces them. She's a dark reddish brown now, except for the webbing between her arms, which is flecked with lichen-like green. Wilson gives her two more fish, which she eagerly accepts. She holds us gently with her suckers as she lets us stroke her head between her eyes. "Nothing else I have ever touched is this soft," I say to Wilson. "Not a kitten's fur, not a chick's down. Nothing is more lovely than this. I could do this all day."

"Yes," Wilson replies, without a trace of sarcasm, "I think you could."

The bliss of stroking an octopus's head is difficult to convey to most people, even to animal lovers. When, back home in New Hampshire, during our walks with our dogs in the woods, I rhapso-dized to my friend Jody, I could tell she was trying hard not to con-clude I had gone insane.

"But," she asked, "aren't they *slimy*? I mean, what about the *slime*?"

It might be more appealing to describe octopuses as slippery. But a banana peel is slippery; slime is a very specialized and essential substance, and there's no denying that octopuses have slime in spades. Almost everyone who lives in the water does. "More of the ocean's residents use, deploy, or are made up of slime than I ever expected," marine scientist Ellen Prager observes. "The undersea world is a seriously slimy place." Slime helps sea animals reduce drag while moving through the water, capture and eat food, keep their skin healthy, escape predators, protect their eggs. Tube worms like Bill's feather dusters secrete slime to build a leathery tube, like a flower stalk, to protect their bodies and keep them attached to a rock or coral. For some fishes—Scott's Amazon discus and cichlids among them—slime is the piscine equivalent of mother's milk. The babies actually feed off the parents' nutritious slime coat, an activity called "glancing." The brightly colored mandarin fish exudes bad-tasting slime to deflect its enemies; the deep-sea vampire squid, an octopus relative, produces glowing slime to startle predators. Bermuda fire worms signal with luminous slime to attract mates like fireflies flashing on a summer night. The female fire worms glow to attract the males; the males then flash, after which the two release eggs and sperm in tandem.

"Kali's and Octavia's slime isn't bad," I told Jody. "Anyway, they're way less slimy than a hagfish."

A creature of the ocean bottom, a hagfish grows to about 17 inches long, and yet, in mere minutes, it can fill *seven buckets* with slime—so much slime it can slip from almost any predator's grip. The hagfish would be in danger of suffocating on its own mucus, except it has learned, like a person with a cold, to blow it out its nose. But sometimes it produces too much slime for even a hagfish to tolerate, and for this occasion, it has devised a nifty trick: the animal wraps its tail around its body like a knot and slides the knot forward, clearing the slime.

"Gross!" Jody cried. "That's so disgusting!" But then she asked me to tell her more about Kali and Octavia's now modest-seeming slime.

Octopus slime is sort of a cross between drool and snot. But in a nice way. And it's very useful. It helps to be slippery if you're squeezing your body in and out of tight places. Slime keeps the octopus moist if it wants to emerge from the water, which some species of octopus do with surprisingly frequency in the wild. Though the infamous "tree octopus" "discovered" in 1998 by researcher Lyle Zapato was a hoax (perpetrated to prove, which it did, that too many young people believe everything they read on the Internet), wild octopuses who live in tidal areas often haul themselves out on land in order to visit different tide pools for better hunting. They may also do this to escape predators in the water, such as another octopus. I had read that in areas blessed by constant ocean spray, an octopus might be able to survive out of water for thirty minutes or more.

"Slime doesn't wreck anything," I explained to Jody. "After all," I reminded her, "slime is part of the two greatest pleasurable experiences known to humankind."

She thought for a moment.

"What's the other one?" she asked.

"Eating," I replied.

❧

"CEPHALOPARTY!" Brendan Walsh's deep voice booms over the hum of the pumps and the heavy metal music playing on the radio. Brendan, thirty-four, tall and burly, works at the aquarium's IMAX theater. Then he goes home to tend to his fish tanks. Right now, he says he has "only" five; he used to have twenty.

He's part of a growing crowd standing around Kali's barrel, waiting for Wilson to open the top so we can play with her. At the aquarium I have joined a cadre of colleagues for whom octopus slime serves as a social lubricant.

Here, too, is Christa Carceo, twenty-five, pretty and petite, with dark hair falling in loose curls down her back, a tiny black jewel perched on its stud above her upper lip, and a smile that lights up the room. "Growing up," she tells me, "other girls had dolls. I had fish." She started with a one-gallon bowl with goldfish, then added bettas, then tetras, guppies, and snails, until she had ten tanks. "You'd go to my room," she says, "all you'd hear is humming." Christa just started volunteering one day a week with Scott in the Freshwater Gallery. She's working as a bartender to pay off her college loans. But what she would really love to do is work at the aquarium.

Marion Britt, the anaconda tamer, returning to the aquarium for the first time since her surgery, has joined our Wonderful Wednesdays, too. With hazel eyes and soft, brown, shoulder-length hair, she has a gentle manner that belies a razor-sharp intelligence, which she applies to her many endeavors, whether devising the first "spot maps" enabling the keepers to tell baby anacondas apart (she sketched their distinctive patterns on pre-drawn forms while holding the foot-long newborns as they bit her) or growing a new exotic yarn business, Purple Okapi, that she can run, despite persistent migraines caused by her surgery, from home.

Today, I also meet Anna Magill-Dohan, who has just completed her sophomore year of high school. Short and small, her dark hair pulled back in a careless ponytail, she has been volunteering at the aquarium for the past two years. During the summer, she puts in four days a week. She has had fish tanks since she got her first as a present at age two. "After that," she tells me, "I kept getting tanks. My parents said no more tanks, but I would just get them and not tell them." Finally she got a pet flounder and her mother found out. As punishment—and here I feared a frying pan would enter the picture, but no—her mother, an elementary school teacher, decreed that she, not Anna, would name the fish. (She named him "Floundie.")

In addition to the usual crowd around Kali's barrel, two aquarium educators have joined us, and Brendan has brought his girlfriend as well. "This is a record," says Wilson. In all, Kali has nine visitors today—more than she has arms for. No other octopus he has known has had this big a backstage fan club.

Though she's never seen this many people before, Kali proves a perfect hostess. She tugs playfully at each arm, looks into our faces, and accepts fish and squid gracefully.

"Wow!" the educators say when Kali's suckers grip their fingers. "Awesome!" whispers Brendan's girlfriend, as a slippery arm coils up to taste her hand.

Around her barrel, we're not only getting to know Kali, and getting her to know us; we are getting to know each other. And for most of us, there is no better way to get to know a person than while petting an octopus. While interacting with Kali, Christa told us about her twin, whose favorite animal is the octopus. Danny has pervasive developmental disorder—a broad diagnostic category for significant, sometimes disabling delay in acquiring basic skills. Christa is working to get legal custody of Danny. Not because her parents, living in nearby Methuen, don't want custody, or because Danny is unhappy there. Vivacious, beautiful Christa wants custody of Danny because, she said, "I can't imagine living without my brother. He wakes up happy every day!"

Danny so loves octopuses that when they go to the aquarium together, he narrates to Christa with great excitement every move the animal makes. "Now she's going up! Now she's moving her arm!"

Once Christa took Danny to a fish market in Boston, and he was upset to find octopus for sale for food. But the carcasses of the slain cephalopods so fascinated him that she eventually bought him one as a present. He keeps it in the freezer and periodically takes it out to look at it.

Thanks to Octavia and Kali, I have begun to learn more about Wilson and his family, too. Born to Jewish Iraqi parents in Rasht, Iran, he grew up attending an American-style Presbyterian mission school in a Persian state, and, at an early age, learned to slip quietly between cultures. When he was sixteen, he went to boarding school in England, and then on to the University of London to study chemistry. He came to America (he remembers the date: January 3, 1957) to study chemical engineering at Columbia University in New York, and moved to Boston to join the Arthur D. Little Corp. There he met his wife, Debbie, a forward-thinking, independent-minded social worker, whose mother was born on the Russian/Polish border and whose dad was American. A year and a half later, Debbie announced that they were going to get married. Wilson instantly agreed. But his conservative, widowed mother was so mortified that Wilson had chosen a woman not of Iraqi Jewish ancestry that she flew to America to try to dissuade him.

Wilson was used to being misunderstood. In a world that demands conformity, in a culture that values animals little, and aquatic animals least of all, we all were. Perhaps that's what helped us bond around a barrel containing a slimy invertebrate that most people considered a monster.

Relatively few people understand, for instance, why Marion would even go into an exhibit filled with giant constricting snakes. "Do you think they *know* you?" people would ask. Of course the snakes knew her, and liked her. And she loved them. Marion wept when Ashley died in summer 2011. Scott understood her feelings perfectly; the moment he'd gotten the call, at 4 a.m. on New Year's Day, that Ashley had given birth, Scott had left his baby son, born just five days before, to rush to the aquarium and tend to the newborn anacondas.

Anna, like all teenagers, feels misunderstood, too. Though a twin, like Christa, Anna is nothing like her athletic, outgoing brother. Extremely smart and forthright, she tells us unabashedly that she is enrolled in a "special" school; that she has Asperger's syndrome, a mild form of autism; that she suffers from migraines, attention deficit disorder, low blood pressure (which once caused her to faint in the anaconda tank), and a tremor; and that she's on various medications. At home, her fish, plus her blue-tongued skink, Laila, help her find some peace; but it wasn't until she started volunteering at the aquarium that she felt truly whole.

"Going behind the scenes at the aquarium changed my life," Anna tells us as we stroke Kali. Before and after sixth grade, Anna spent part of her summers at "fish camp" at the aquarium. Then when she was fourteen, she started taking an art class on Saturdays, and after class would take the T to spend the rest of the day at the aquarium. Dave Wedge, the bearded, outgoing former high school teacher who runs the Edge of the Sea exhibit and the Education Center's Wet Lab, recognized her from fish camp and invited her to see his lab. He told her to meet him in an hour. But Anna had no sense of time and did not own, and could not read, an analog watch. So she waited outside the door to the Wet Lab for an hour—in the pouring rain. Dave was so impressed that, even though Anna was too young to be an official volunteer, he started finding things for her to do behind the scenes.

Now an official volunteer, Anna not only has a digital watch (and knows how to read it), but she also knows the common and Latin names for every marine vertebrate and invertebrate in the aquarium. She apologizes that she hasn't memorized all the ones in Freshwater yet.

"The people here are as different from regular people as an octopus. I feel at home here," Anna says, speaking for all of us, "like I belong."

Belonging to a group is one of humankind's deepest desires. We're a social species, like our primate ancestors. Evolutionary biologists suggest that keeping track of our many social relationships over our long lives was one of the factors driving the evolution of the human brain. In fact, intelligence itself is most often associated with similarly social and long-lived creatures, like chimps, elephants, parrots, and whales.

But octopuses represent the opposite end of this spectrum. They are famously short-lived, and most do not appear to be social. There are intriguing exceptions: Male and female lesser Pacific striped octopuses, for instance, sometimes cohabit in pairs, sharing a single den. Groups of these octopuses may live in associations of forty or more animals—a fact so unexpected that it was disbelieved and unpublished for thirty years, until Richard Ross of the Steinhart Aquarium recently raised the long-forgotten species in his home lab. But the giant Pacific, at least, is thought to seek company only at the end of its life, to mate. And even that is an iffy proposition, as one known outcome is the literal dinner date, when one octopus eats the other. If not to interact with fellow octopuses, what is their intelligence for? If octopuses don't interact with each other, why would they want to interact with us?

Jennifer, the octopus psychologist, says, "The same thing that got them their smarts isn't the same thing that got us our smarts." Octopus and human intelligence evolved separately and for different reasons. She believes the event driving the octopus toward intelligence was the loss of the ancestral shell. Losing the shell freed the animal for mobility. An octopus, unlike a clam, does not have to wait for food to find it; the octopus can hunt like a tiger. And while most octopuses love crab best, a single octopus may hunt many dozens of different prey species, each of which demands a different hunting strategy, a different skill set, a different set of decisions to make and modify. Will you camouflage yourself for a stalk-and-ambush attack? Shoot through the sea with your siphon for a quick chase? Crawl out of the water to capture escaping prey?

Losing the shell entailed a trade-off: Now that the animal is "a big packet of unprotected protein," as one researcher put it, just about anything big enough to eat it will do so. Octopuses are well aware of their vulnerability and make plans to protect themselves. Jennifer saw this clearly when she was watching a common octopus in Bermuda on an expedition in the 1980s. Returning home from a hunting expedition, the octopus was clearing the front of the den with its arms. Then, suddenly, it left the den, crawled a meter away, picked up a rock, and placed it in front of the den. Two minutes later, the octopus ventured forth again to select a second rock, and then a third. Attaching its suckers to both rocks, it lugged the load home, slid through its den opening, and then carefully arranged the rocks in front of the lair like a stone fortress in front of a castle. What the octopus was thinking seemed obvious, Jennifer said: "'Three rocks are enough. Good night!'" Now it felt safe enough to go to sleep.

In 2009, researchers in Indonesia documented octopuses that were carrying around pairs of half coconut shells, which they used as portable Quonset huts. With obvious effort, the octopuses would lug the shell halves, nested one inside the other, beneath their bodies as they walked stiff-armed across the sandy bottom, then assemble the half shells into a sphere and climb inside. At the Middlebury octopus lab, assistant animal caretaker Caroline Clarkson noticed another instance of tool use. A sea urchin was feeding too near the entrance of the den belonging to a female California two-spot. So the octopus ventured out of her lair to pick up a 3.5 by 3.5-inch piece of flat slate lying six inches away and dragged it back to the den, where she erected it like a shield to protect herself from the urchin's spines.

From building shelters to shooting ink to changing color, the vulnerable octopus must be ready to outwit dozens of species of animals, some of which it pursues, others it must escape. How do you plan for so many possibilities? Doing so demands, to some degree, anticipating the actions—in other words, imagining the minds—of other individuals.

The ability to ascribe thoughts to others, thoughts that might differ from our own, is a sophisticated cognitive skill, known as "theory of mind." Once it was thought to be unique to humans. In typical children, theory of mind is believed to emerge around age three or four. The classic experiment goes like this: A toddler views a video of a girl who leaves a box of candy behind in her room. While she's gone, an adult replaces the candy in the box with pencils. Now the child comes back to open up her box again. The experimenter asks the tot, what does the little girl expect to find in the box? The toddler will say: pencils. Only an older child will understand that the little girl would *expect* to find candy, even though that's not what's really there.

Theory of mind is considered an important component of consciousness, because it implies self-awareness. (*I* think *this*, but *you* might think *that*.) Dr. Brian Hare, director of the Duke Canine Cognition Center, recently demonstrated that dogs understand that others might have knowledge that they do not possess. As an experiment, he presented dogs with two smell-proof containers, one with food, one without. The dogs quickly figured out that the people knew what they did not, and would follow a human's pointing finger to the hidden treats.

This is precisely what Nancy King's octopus, Ollie, was doing when she followed her finger to discover the crab she couldn't find by herself.

Of course, there are many other examples. The birds of prey with whom falconers hunt look to the falconer, or to her dogs, to flush game. African honey badgers follow certain birds (known as honey guides) to find bees' nests. Both parties seem to realize that when badgers open up the nests to eat the honey, the birds can then feast on the bee larvae.

But of all the creatures on the planet who imagine what is in another creature's mind, the one that must do so best might well be the octopus—because without this ability, the octopus could not perpetrate its many self-preserving deceptions. An octopus must convince many species of predators and prey that it is really something else. Look! I'm a blob of ink. No, I'm a coral. No, I'm a rock! The octopus must assess whether the other animal believes its ruse or not, and if not, try something different. In Jennifer's book, she and her coauthors report that specific displays are directed at particular species under specific conditions. The Passing Cloud display, for instance, is used by an octopus to scare an immobile crab into moving and thus giving itself away. But to fool a hungry fish, an octopus is more likely to use a different strategy: to rapidly change color, pattern, and shape. Most fish have excellent visual memories for particular search images, but if the octopus changes from dark to pale, jets away, and then turns on stripes or spots, the fish can't keep track of it.

To survive long enough to meet us at the New England Aquarium, Kali may have met and matched wits with many different species of bird, whale, seal, sea lion, shark, crab, fish, and turtle, as well as other octopuses and human divers—all with different kinds of eyes, different lifestyles, different senses, different motives, different personalities, and different moods. Compared with most people, whose daily lives involve direct interaction with only one species, Kali is a cosmopolitan sophisticate, and we are small-town bumpkins.

And right now, she's working the crowd. Kali is curious about her company—and what is more endearing than someone showing interest in you? She explores Brendan and his girlfriend with the tip of her second left arm while she investigates the two educators by folding a sucker around their fingertips. She flips upside down, unfurling the creamy suckers on her arms like a blooming flower. Christa, Anna, Marion, and I offer our hands and forearms; she attaches her suckers and pulls gently, seemingly playful. Her skin mottles; she creates thorns and horns; she pulls her head up and lets me pet it again, now going white beneath my touch. Her eye rolls. She is looking for Wilson. She finds his face, and two of her arms rise up and envelop his arm like two slices of bread around sandwich filling.

Bill, watching the scene from behind us, is delighted. Kali is active, interested, friendly, and outgoing. "She's going to be a wonderful octopus for display," he says proudly.

❊

Even though it's not a Wednesday, Wilson and I have made a special trip to the aquarium. Today we are celebrating Christa and Danny's birthday. With the cooperation of Bill and Scott, we're here to share Christa's surprise for her brother.

Last night, Danny took the bus from their parents' home in Methuen to Christa's apartment in Boston. At 11:15 a.m. Wilson and I are waiting, ready for Christa to bring her brother behind the scenes on the third floor.

"He was always reading encyclopedias," she boasts. "My sister and I would glance at them, but he would read them. My mother ended up getting a lot of encyclopedias," she said. Danny's favorite entry, since he was thirteen, was the octopus. What about octopuses most fascinates him? "Their appearance," he says. "How smart they are. They're covered with suction cups!"

Last night, Christa says, she read my article in *Orion* to Danny. She whispers conspiratorially to me, "He said, 'Can you imagine getting to touch an octopus?'" All he knew yesterday about today's plan was that they were going to the aquarium together. "So we'll see the octopus today," he had said to her this morning. "It's going to be a good day."

He has no idea what we have in store.

Wilson leads Danny over to Octavia's tank. "Guess whose tank this is?" Christa asks him.

Danny's eyes widen. "The Big O?"

Wilson attempts to attract Octavia, proffering a fish in the tongs. Christa, Danny, and I rush downstairs to the public viewing area to see how Octavia reacts. Danny waves at her through the glass. Octavia ignores the tongs at first. Finally she grabs them with two arms, three arms—and turns bright red. The fish drops. She doesn't want to eat. She lets go of the tongs and Wilson withdraws them.

Wilson appears in front of the tank with us. "Did he see that?"

"That was amazing!" says Danny. That was surprise enough for him. But then we go back upstairs and stand by Kali's pickle barrel. Wilson starts unscrewing the lid.

"Hey, Danny, check this out," Christa says as Kali, dark reddish brown, floats to the surface.

"I always thought there was just one octopus in this place!" Danny says. Wilson extends his hand and Kali covers it with her suckers.

Danny begins to shake with excitement. "Here, give her a fish," Wilson says to him. "Put it in the sucker and let her take it," he urges.

Danny holds the fish, but he is leery at first. "I think she's grabbing!"

"Let it go—let her take it," says Wilson. "She won't hurt you. Put your hand in the water!"

Kali's head and three of her arms are now out of the water, coming up over the edge of the tank: She is eager to greet us. We're all petting her and urging Danny to do the same. But he's frightened. He pokes at a single sucker with one finger and withdraws, shaking. He can't help it: Later he tells me he was thinking of a TV show in which he saw an octopus as big as a building attacking people.

Suddenly, a fountain of water gushes up from the barrel. "That's her saying hi to you!" says Christa. This is followed by another gusher, and then a much taller spout—hosing Danny right in the face.

This doesn't bother him a bit. He looks no more or less dazed than he did before. He is in the dazzling presence of an octopus, thrilling and frightening at the same time.

Dripping, Danny reaches a finger out to touch one of Kali's suckers.

"I do have a frozen octopus in my freezer," he says to me, "but that one's dead."

Kali starts to heave her gelatinous mass out of the tank toward us. "Here she comes!" says Christa. Wilson and I try to urge some of her arms back in the water. Kali attaches her suckers to our arms. "She's much more eager to touch me than him," Wilson says to me. "It's the nervousness. She can feel his nervousness. I've never seen it more clearly than that."

"If you were a crab or a fish," Wilson tells Danny, "she'd move you all the way down to her mouth. But you're a human so she won't." Instead he hands Danny another fish. "Let it go. She'll take it."

And she does.

"Oh, this is awesome!" says Danny. He waves at her, wiggling the fingers of his left hand.

And now he feels safe enough to give her his hand. Kali gently attaches five suckers, then ten, and now perhaps twenty to the palm of his hand. "She feels like a rubber glove!" Danny says.

"His nervousness is decreasing and she's more willing to interact," says Wilson. "She has much more awareness of us than we do of her."

"I think she really likes me!" Danny says to us, astonished.

"Her name is Kali," Christa says.

"Hi, Kali," Danny says, as if to a person. She is moving, sucker by sucker, up the side of the pickle barrel, rolling forward like a Slinky.

But now Wilson feels we might be in danger of exhausting her. He puts the top back on.

Danny is starstruck. "I petted a live octopus at the aquarium!" he cries. "Wow, that was adventurous! I can't wait to tell my parents! She liked me, too!"

And there's more. Now Wilson brings out a jar and removes the blue surgical glove covering the top. Inside, about an inch long, black, chitinous, and in two interlocking, curved pieces, is one of Wilson's prize possessions.

"Do you know what this is?" he asks Danny.

"A shell?"

"No—"

Danny remembers a picture from the encyclopedia. "It looks like an octopus beak!"

"This was the beak of a very old octopus," says Wilson. It was George's beak. "And it's for you."

Danny is stunned.

"What do you think?" asks Christa.

"It came from a real octopus!"

Wilson has brought another gift for Danny from his collection: a mounted photo of George by photographer Jeffrey Tillman. "I'll put it in my room," says Danny, awed. "All I have to do is put in a nail. I'll put it right by my bed."

Danny and Christa and I will spend the rest of our day together at the aquarium, but Wilson has to leave early. He got the call earlier this morning: There is a bed for his wife in a nearby hospice. If she is going to take it, she must move today. The doctors still don't understand what is wrong with her, only that her self and her strength are ebbing away, and there seems no stopping it. Wilson's afternoon will be spent getting his wife, with whom he's traveled the world, ready for her final journey. He is making plans himself to move from their large, beautiful home, with its huge kitchen with tiles around the stove and many bedrooms for visiting guests and grandchildren and Debbie's home office. Preparing to move to a smaller place, Wilson is giving away treasures—he has given Christa and Marion and me coral and shells and books, and donated large specimens to the aquarium. And yet, in the face of looming tragedy, Wilson has chosen to be with us this morning, celebrating the birthdays of these two young, happy people.

He has managed to make this day a good one—a miracle of sorts. And who better to preside over such a miracle than an octopus, wielder of otherworldly powers—an octopus named after Kali, the goddess of creative destruction, the deity embodying the opposites of kindness and cruelty, sorrow and joy?

❦

A bright summer afternoon in Boston: Outside, park rangers in hats answer questions about the whale watch and harbor view cruises, happy children and parents spin and shout on the merry-go-round on the Common, while grownups crowd Faneuil Hall eating soft pretzels and ice cream. Inside the aquarium, Anna is assisting Scott, Christa is distributing black worms, and Bill is feeding the endangered freshwater turtles, red-bellied cooters whom he is raising for the state of Massachusetts for release. Wilson and I are with Kali, who has finished eating her squid. Still upside down, she lingers at the surface. She holds one of my fingertips in one of her suckers and squeezes it periodically, like you'd squeeze a hand you were already holding. One of her arms encircles Wilson's wrist; another holds his other hand and forearm. I reach with my free hand to her head and begin to stroke her.

From all appearances, the three of us are as languid as the summer day, as if time has eddied in its flow and we have escaped the confines of clock and calendar, and perhaps even species. "If someone came upon us now," I say to Wilson, "they might think we're members of some strange religious cult."

"Cult of the octopus?" Wilson chuckles softly.

"The route to peace and ecstasy," I answer.

"Yes," says Wilson, his voice soft as a lullaby. "This is very peaceful."

While stroking an octopus, it is easy to fall into reverie. To share such a moment of deep tranquility with another being, especially one as different from us as the octopus, is a humbling privilege. It's a shared sweetness, a gentle miracle, an uplink to universal consciousness—the notion, first advanced by pre-Socratic Greek philosopher Anaxagoras in 480 BC, of sharing an intelligence that animates and organizes all life. The idea of universal consciousness suffuses both Western and Eastern thought and philosophy, from the "collective unconscious" of psychologist Carl Jung, to unified field theory, to the investigations of the Institute of Noetic Sciences founded by Apollo 14 astronaut Edgar Mitchell in 1973. Though some of the Methodist ministers of my youth might be appalled, I feel blessed by the thought of sharing with an octopus what one website (loveandabove.com) calls "an infinite, eternal ocean of intelligent energy."

Who would know more about the infinite, eternal ocean than an octopus? And what could be more deeply calming than being cradled in its arms, surrounded by the water from which life itself arose? As Wilson and I pet Kali's soft head on this summer afternoon, I think of Paul the Apostle's letter to the Philippians about the power of the "peace that passeth understanding . . ."

And then—SPLASH!—we're hosed.

Kali's funnel, less than an inch in diameter, manages to hit us both at once, soaking our faces, our hair, our shirts, and our pants with 47°F salt water.

"Why . . . ?!" I sputter. "Is she mad at us?"

"That was not aggression," says Wilson. We both lean over the barrel and see she has sunk to the bottom, from where she looks up at us innocently. "That was playful," he says. "Remember, they are all individuals." We put our hands right back in. But she doesn't attach her suckers right away. Instead, she points her funnel at us, like a kid aiming a squirt gun. I'm not fast enough to dodge it, but I can't help watching to see what she does next. She rises so that her head now lies just beneath the surface, and I see the water swelling from the pressure of her siphon. Clearly, she can modulate the flow with great precision.

She can also move the funnel with astounding flexibility. I had assumed the organ, though supple, was firmly attached to just one side of her head. But Kali shows us that clearly it is not. At one moment, her funnel is on the left side; the next, she's swung it 180 degrees, to her right. It's as surprising as if you saw a person stick his tongue out his mouth, and next out his ear—and then out his other ear.

Then Kali fluffs up the suckers on her arms like the frills on a petticoat and waves her arms at us. If she were a person, we could reach no other conclusion than that she is teasing us, coyly daring us to try again.

When it's time for me to tear myself away, I go down the hall to say goodbye to Scott in the Freshwater Gallery. Earlier that day, I had apologized for causing inconvenience, since he always has to send someone downstairs to the lobby to get me every time I show up at the aquarium, to take me behind the scenes. The few times I came up unaccompanied, I was stopped by staff worried about thieves. (The items most often stolen—before locks were installed on their tank tops—were small turtles like Bill's red-bellied cooters.) So Scott has spoken on my behalf to Will Malan, one of the coordinators of the aquarium's extensive volunteer program. Six hundred and sixty-two adult volunteers donate an estimated $2 million worth of time to the aquarium, performing tasks from cleaning up penguin poop, to giving educational talks, to feeding and moving animals to helping to design new exhibits. Another hundred young people help through internships and teen volunteer programs. All wear badges identifying them as volunteers, allowing them behind the scenes.

I fit none of these categories, but Scott ushers me into Will's office, where Will takes a photo that will be emblazoned on my new badge. Half my hair is still plastered to my head from Kali's dousing, but I am overjoyed with the title Will and Scott have given me: I am now the aquarium's official "Octopus Observer."

The badge is a talisman. It gives me access throughout the aquarium, even outside of public visiting hours. This will prove indispensable, for now I have yet another reason for visiting the aquarium:

Octavia has laid eggs.

Eggs

Beginning, Ending, and Shape-Shifting

Because she has retreated to the far corner of her den, beneath a rocky overhang, I can see Octavia only from the public viewing area now. In the summertime, the New England Aquarium averages six thousand visitors a day, so in order to beat the Boston commuter traffic and arrive before the aquarium opens to the public, to watch Octavia in peace, I rise at 5 a.m. and start driving.

I park in the aquarium garage in the coveted Crab section on the third floor. (If I arrived after 9, I would be consigned to Jellyfish on the fifth.) Entering the aquarium, I wave to the staff manning the information desk and begin my spiraling walk up the incline: past the penguin pools with its noisy little blues, Africans, and southern rockhoppers; past the Blue Hole exhibit with its Goliath groupers; past the Ancient Fishes, the long, silvery arowana with its bony tongue, and the primitive lungfish with its curious, lobe-like fins; and past the Mangrove Swamp. I tread beneath the hanging skeleton of the North Atlantic right whale, pause to greet the electric eel, briefly watch the trout, and head to the Gulf of Maine exhibit, the Isles of Shoals tank, and the three-foot-long, flat, lumpy, bottom-dwelling goosefish. Just past the Pacific Tidepool, right before the "staff only" stairway to the Cold Marine and Freshwater galleries and the elevator to the top of the Giant Ocean Tank, my step grows quicker, my heart beats faster—for next, I come to the tank where I will see my friend Octavia.

She appears to be sleeping, plastered to the roof of her lair. Her skin texture and color are almost indistinguishable from the rock, her pendulous head and mantle hanging upside down. Her left eye is open but her pupil is a hair-thin slit. Her right eye is obscured by the thick part of one arm, its suckers facing me, until the arm curves backward, out of sight. The tips of five of her arms hang in curling tendrils from the roof and sides of her lair. I can't see her gills or any sign of breathing. The movements of her body appear to be due only to the currents in the water.

I stand transfixed before her tank, watching her with the red covering on my headlamp so as not to disturb her with the light. Visiting her at this early hour, before the dim bulb of her exhibit is turned on, is like a meditation for me. I must prepare my senses, let my eyes adjust to the dark. This demands patience. I must train my brain to switch from seeing nothing at all to seeing subtle changes, to recognizing that suddenly, a great deal might be happening, all at once.

Now, Octavia is the picture of peace, an octopus Madonna. She seems to have grown larger since I last saw her. Her head and mantle are the size of a watermelon you'd bring to a family picnic. She cradles something in the webbing between some of her arms; I can see the weight of what she holds, but not the content. Like a sleeping person, she occasionally stretches some of her arms, but otherwise she is still.

Then, at 9:05, seventy-eight minutes after my arrival, she starts moving. Her body begins beating like a heart. She fills her gills with deep breaths of salt water and shoots them out her funnel. She moves one arm across her body, almost absentmindedly, like a pregnant woman caressing her big belly. Two of her other arms rub each other, cleaning the suckers. And with these movements, Octavia reveals some of the treasures she is guarding. A two-inch chain of about forty eggs, each the size and color of a grain of rice, pops into view. It hangs from the ceiling of her lair and drapes over one of her arms like an errant lock of hair sweeping over a woman's shoulder. These eggs are the hidden treasures resting earlier in the webbing of her arms.

There are many more eggs here than I can see. Some clusters are as many as eight inches long. They are stacked five or six clusters deep toward the back recess of her den. But her body is now covering almost all of them.

This is why Octavia no longer wants to interact with us. She has more important things to do. Caring for her eggs is the job that will last a female octopus to the end of her life.

Octavia started laying eggs in June, while I was in Africa, but no one has seen her lay a single one. "You go in first thing in the morning, and there are more eggs," Bill says. Giant Pacific octopuses are generally nocturnal, and certainly a process as delicate as egg laying is best undertaken beneath the dark's safe shroud. Unseen, Octavia has been crawling to the roof of her den to pass each tiny, teardrop-shaped egg out her siphon. Each egg's narrow end has a short cord attached. Using several of the smallest suckers nearest her mouth, she carefully weaves between thirty and two hundred eggs into a string, like one might braid onions. Using a secretion from glands inside her body, she glues the clusters to the roof and sides of her den, where they hang like bunches of grapes. And then she begins another chain, and another. In the wild, over the course of about three weeks, a female giant Pacific octopus might lay between 67,000 and 100,000 eggs.

It is very unlikely that Octavia's eggs are fertile. Female octopuses store inactivated sperm in their spermatophoric gland for months, then activate it when it's time to fertilize their eggs; but Octavia would have had to have mated before her capture, more than a year ago. She was probably then too young to have accepted a spermatophore from a male.

And yet, Bill is visibly proud of Octavia's eggs. Even though egg laying signals the approach of the end of a female's life, Bill is not sad. He seems increasingly content, satisfied with each new egg chain that appears. For him, the process represents completion.

"With Athena, I felt gypped that she died so early," he says. The proper end of any female octopus's life *should* be laying eggs. Guarding them, aerating them, cleaning them, Octavia will be able to complete the same rituals her mother completed before her, and her mother before her, and her mother before her, back hundreds of millions of years.

In her memoir of living among the Bushmen, *The Old Way: A Story of the First People*, my friend Liz lovingly invokes an image first coined by evolutionary biologist Richard Dawkins: "You are standing beside your mother, holding her hand. She is holding her mother's hand, who is holding her mother's hand. . . ." Eventually the line stretches three hundred miles long and goes back five million years, and the clasping hand of the ancestor looks like that of a chimpanzee. I loved picturing one of Octavia's arms stretching out to meet one of her mother's arms, and one of her mother's mother's arms, and her mother's mother's mother's. . . . Suckered, elastic arms, reaching back through time: an octopus chorus line stretching not just hundreds, but many thousands of miles long. Back past the Cenozoic, the time when our ancestors descended from the trees; back past the Mesozoic, when dinosaurs ruled the land; back past the Permian and the rise of the ancestors of the mammals; back, past the Carboniferous's coal-forming swamp forests; back past the Devonian, when amphibians emerged from the water; back past the

Silurian, when plants first took root on land—all the way to the Ordovician, to a time before the advent of wings or knees or lungs, before the fishes had bony jaws, before blood pumped from a multi-chambered heart. More than 500 million years ago, the tides would have been stronger, the days shorter, the year longer, and the air too high in carbon dioxide for mammals or birds to breathe. All the earth's continents huddled in the Southern Hemisphere. And yet still, the arm of Octavia's ancestor, sensitive, suckered, and supple, would have been recognizable as one of an octopus.

In the wild, most female octopuses lay eggs only once, and then guard them so assiduously they won't leave them even to hunt for food. The mother starves herself for the rest of her life. A deep-sea species holds the record for this feat, surviving four and a half years without feeding while brooding her eggs near the bottom of Monterey Canyon, nearly a mile below the surface of the ocean.

At the Octopus Symposium in Seattle, scuba diver Guy Becken had given a PowerPoint presentation about Olive, a wild giant Pacific octopus who lived only a mile across the water from the Seattle Aquarium, at a popular dive site known as Cove 2. As a member of the local Tuesday-night divers' club, he and his friends would visit the area and often find octopus there, as well as six-gill sharks, wolf eels, and lingcod. In 2001, they regularly encountered a large male octopus they called Popeye just 100 feet from shore among the pier pilings. Then in February of 2002, another octopus appeared, who proved to be a female. They estimated she weighed about 60 pounds. They named her Olive.

Olive grew so accustomed to the divers that she would accept herring they handed to the suckers on her outstretched arm. By late February, though, she would not leave her den. Under a cluster of sunken wooden pilings, she created a fence of eight-inch rocks in a semicircle in front of one the den's two openings. But the divers could still see inside, and by the end of the month they confirmed that Olive had laid eggs.

"Each trip to see her was different," Becken said. "Sometimes she was receptive. Sometimes she clearly didn't want the company." During the first month after she laid her eggs, she would still take the herring the divers offered. But "then," Becken said, "she'd start throwing the herring back."

Hundreds of divers came to see Olive on her eggs that summer. Rapt, they watched her caressing the eggs with her suckers and blowing water through them with her siphon; they saw her fend off sunflower sea stars who tried to investigate her brood chamber, hungry for her eggs. By mid-June, the divers could see the babies' black eye spots developing inside the eggs. "There's one! There's one!" Becken said, his excitement still fresh years later as he pointed out the developing eyes of the unborn babies to us on the projection screen.

On a night dive in late September, Becken and his friends witnessed some of the first of Olive's offspring (called paralarvae) emerging from the eggs. Olive used her siphon to blow the newborn, tiny, perfect octopuses, each no bigger than a grain of rice, out of the eggs, out through the den opening. There, they would float away on the current like the spiderlings ballooning on air at the end of *Charlotte's Web*. Until the survivors grew large enough to settle on the bottom, they would become part of the ocean's wandering plankton, the floating mix of millions of tiny plants and animals that form the basis of the food chain, make most of the world's oxygen, and keep the world alive.

The development of octopus eggs is at least partially dependent upon temperature. Off California's coast, the eggs of giant Pacific octopuses typically take four months to hatch; in Alaska's colder waters, seven or eight. Olive's took longer than the six months that is usual in Puget Sound, and the last of her babies hatched in early November. Just days later, divers found her corpse, milky and translucent as a ghost, just outside the den. Two sea stars were feeding on her body.

"It was sad," Becken said. "Some divers didn't want to look. But since her life, since her death, this place has always been known as Olive's Den. And very seldom do you go out there and not find an octopus. Whenever we see them," he said, "we think of Olive's legacy."

Here, in our aquarium on the other side of the continent, Octavia still accepts the fish Bill and Wilson hand her at the end of the grabber. "This means she might live for several more months," Bill assures me.

During these months, Octavia will give us a close-up and detailed view of her most intimate, ultimate task, far more than would be possible to see in the wild. Her attentions will not be able to transform her infertile eggs into living paralarvae. But other transformations will be revealed around her tank, sometimes thanks to Octavia herself—some of them sad, some of them strange, and some, like Octavia's eggs, a whispered promise of new life.

⁂

"She's still strong," Wilson says with relief, feeling the pull of Octavia's arms on the tongs as he hands her a squid. "She's got a while to go yet."

Kali, meanwhile, is growing larger, stronger, and bolder by the day. Anna already has both her hands in the sump and is communing with the tips of Kali's arms as they protrude from the holes in the pickle barrel. When Wilson unscrews the top, Kali floats up to look at him immediately. We all plunge in our hands. Kali flips upside down to receive two capelin eagerly in separate arms, and starts passing them from sucker to sucker toward her mouth. Meanwhile her other arms are busy with ours. We feel like we're taking a cold bath in suckers.

We've been together like this for only about three minutes when Kali looses a water bomb. We all feel it, but Anna is hit directly in the face. She's soaked. Freezing salt water drips from her dark hair and from the tip of her nose. A full second later, Anna yells: "AAAHHH!"

It takes a moment for us to understand what has happened. At first, we assume her yell is a delayed reaction to the soaking. But then we see that three of Kali's arms have closed like a Venus flytrap around Anna's left arm. We rush to peel the suckers away, each cup making a loud pop as the suction is broken. Anna steps back from the tank, with impressive calm, to examine her left hand. Just at the lowest joint of the thumb there are two dents bearing the imprint of Kali's upper and lower mandibles.

Marion helps her rinse off the wound at the sink. Though the skin is broken, no blood yet wells from the punctures. But this may be because of Anna's low blood pressure.

Anna is not in pain, nor is she frightened. But the rest of us are alarmed. Christa, hearing the commotion from the hallway, rushes in, and helps Wilson and me as we try to stuff the octopus back in her tank to put the lid on. It's not an easy job. As soon as Kali sees the lid coming, she starts scrambling out, rising over the top like the foaming head on a beer. Her arms grip the lip and sides of the barrel as fast as our six hands can peel her suckers away. I feel bad about ending our interaction so soon; this is certainly the most interesting part of Kali's day, and it seems clear she doesn't want it to end.

But we need to attend to Anna. Almost instantly, two of the aquarium's team of first aid workers appear at her side. They were on the floor when the bite happened. *Now* Anna is nervous. She doesn't want this to turn into a big deal; she doesn't want to get in trouble. Most of all, she doesn't want to be banned from interacting with Kali.

The first aid responders are worried. Even though the wound is tiny, less dramatic-looking than a nip from a parakeet, this is an octopus bite, and there hasn't been an instance of this in nearly a decade, since Guinevere bit Bill. "Do you feel dizzy?" they ask Anna. Though the giant Pacific's is among the least toxic octopus venom to humans, still, envenomated wounds can take many weeks to heal; plus there is the possibility of an allergic reaction, like some people have to bee stings. "Do you have a burning sensation?" they ask. Anna does not. Soon it's clear that, although Kali could have chosen to chomp down and inject venom, instead she merely gave Anna a little pinch. Anna is fine.

Yet Wilson is horrified. "She is being aggressive!" he says in astonishment. "I've had hundreds of interactions with octopus. My granddaughter interacted with an octopus when she was just three years old!" And Kali is one of the sweetest, most outgoing octopuses he has known, the octopus who has interacted with humans far more frequently than any of her predecessors.

What was going on? Might Kali have mistaken Anna's hand for a fish? Such an error is unlikely. Even our clumsy fingers, bereft of chemoreceptors, can tell the difference between a person's skin and a fish's slime-covered scales. Did Kali bite Anna randomly? She could have bitten any of us just as easily; all our hands were in the tank. But with her funnel's perfect aim, she purposely targeted Anna's face, right before biting her. The bite was intended for Anna and Anna alone. Why would she nip this gentle, smart, loving, experienced teen?

I wonder if it's because of Anna's tremor. Kali had squirted Danny, too, when he was shaking. But more likely, I wonder about

Anna's medication. She's on several different meds, and her doctors change these frequently. Perhaps Kali could taste them, and the switch confused her. Maybe Anna doesn't taste like her usual self today. And in fact, Anna tells me, her doctors did recently change a prescription.

We head for an early lunch, to reassure Anna she's done nothing wrong. We trade stories about various species that have bitten us. The anaconda, Kathleen, bit Scott when he was holding her for an X-ray. (No reptile likes contact with a cold metal table.) I had been the toast of my aerobics class at home when I turned up with a bandage on my hand, whose origin was a bite from an arowana, a predatory Amazon fish who had leapt from the water of its tank to bite me when I was feeding it behind the scenes. Anna had already been bitten by a remarkable number of animals—including a piranha (whom she was removing from a hook on a trip to Brazil with Scott's sustainable fisheries organization), a small shark at the aquarium, and, more surprisingly, a chicken. She seems rather pleased to add octopus to her list.

"There's a certain thrill to being bitten," says Christa. This might not be true for most people, but those of us gathered at our lunch table agree: Being bitten is an intimate interaction—and often, especially with marine creatures, one that carries no malice. Even "attacks" on humans by great white sharks are thought to be exploratory, not predatory. This might well have been the case with Kali and Anna.

Behind the scenes in the Freshwater Gallery, one of the young volunteers has drawn a cartoon of an electric eel in black magic marker on the wall, with a thunderbolt emanating from the head and the words *Have you tried it?* And yes, I, too, experienced the thrill of 600 volts coursing from Thor, the electric eel behind the scenes (the one on display is named Mittens), through my skin when

I purposely touched the soft, slippery back of his head. ("Use your right hand," Scott advised me jokingly, "since the left is closer to your heart.") It feels like putting your finger in an electrical outlet. Being shocked by the electric eel is sort of an initiation into an exclusive club.

Some of this is just plain fish-nerd machismo. Still, most bites are accidents, and we all know most accidents happen when you're being lazy or sloppy, which is nothing to be proud of. Yet a bite is proof of a kind of contact that—even when it goes wrong—at a time when most people are increasingly isolated from the natural world, we are privileged to experience. Though the denizens of the aquarium live in captivity, they are still, at heart, wild animals. A bite from a fish or an octopus is proof we are willing, even eager, to literally give ourselves (even tiny, actual pieces) to the animals here, in order to touch the wild.

⁂

During the summer of Octavia's eggs, I find transformation everywhere I look.

Octopuses are masters of change. One day, I find Octavia white as a sheet—a color I have only seen on her in patches before. White is the color to which an octopus increasingly reverts when it gets older, as the muscles controlling the color-producing chromatophores lose tone with age. Another day, I find that the tip of Octavia's third right arm, R3, is missing. Was this always the case and yet none of us, mesmerized by all those arms, always in motion, ever noticed before? Julie Kalupa, a diver and medical student at the University of Wisconsin, writes that a giant Pacific octopus can regenerate up to one third of a lost arm in as little as six weeks. Unlike a lizard's regenerated tail, which is invariably of poorer quality than the original, the regrown arm of an octopus is as good as new, complete with nerves, muscles, chromatophores, and perfect, virgin suckers. Even the specialized arm of the male, the ligula, can be regrown (though this reportedly takes a while longer).

Kali, too, continues to astound. One day, we discover she is training us. Christa, Marion, and Anna join Wilson and me as he unscrews the lid of her barrel. Kali is already at the top, reddish brown, looking at us with curious, animated eyes. The moment the lid is off, two arms, three arms, five arms, then her whole body comes heaving out of the tank. Her suckers are eager to grab us and anything else she can reach. We gently detach her suckers from the outside of the barrel, hoping she will be content to play with us instead of trying to escape. Her twisting arms explore our hands for a moment, but then she sinks and flips upside down, like a frustrated child flops down for a tantrum. Then she floats up, suckers first, hovers at the top for a moment—and then, like an upside-down umbrella opening, expands her body. Before we can see her funnel swing and aim at us, she lets loose a torrent.

We women see our pants and shoes are wet, but it's only Wilson—the person she likes the best, the person who usually hands her the first fish of the day—who is roundly hosed. "That was meant for me," says Wilson, his face dripping. "She is going to be a handful!"

Why did she squirt this time? Is she annoyed we pushed her back in her tank when she wanted to explore? Is she playing?

I sense it's something else. I suspect she is holding us up for capelin at gunpoint—a squirt gun, that is. "I think she wants you to give her a fish," I suggest, "and she wants it first thing."

The dish of fish is only an octopus-arm's length away from her barrel, and Wilson grabs a capelin. He hands it to the suckers of one of her arms. Then Christa places a second fish in the pillowy, white cups of another arm. Instantly Kali becomes exceptionally calm. Lying upside down at the surface, arms splayed, she gives us an extraordinary view of her shiny, black beak. This is the first time even Wilson has seen the beak inside a living octopus. It is a private and trusting moment, her sharing with us this surprising part of her, normally hidden inside at the confluence of her arms. We watch the first fish pass from one sucker to another, tail first. The three-inch capelin disappears in ten seconds. The second fish is eaten a bit more slowly. Its pink innards squish out from the chewing motion of Kali's beak, but then the fish slides slowly inside her . . . the silvery eye, and then the top of the head, gone.

After that, we always greet Kali immediately with a fish or squid. The hosing stops for the entire summer.

Kali now gets many visitors these days besides us—perhaps too many, Wilson worries. He fears she may be overstimulated and closes the lid on the barrel.

At times like this, after Kali has been fed, when the front of Octavia's tank is too crowded for me to observe her, and there's nothing pressing going on in Cold Marine or Freshwater, Anna, Christa, and I cruise the other tanks of the aquarium, roaming like girls window-shopping downtown. But to us, each tank is more like a station of the cross, a site for a series of devotions. Here we are sanctified, baptized over and over by the beauty and strangeness of the ocean.

In the tank two down from Octavia's, a 60-foot-long veil of gossamer, studded with pearls and diamonds, floats at the surface above the flattish, three-foot-long goosefish, an animal the color and texture of bottom detritus, whose large mouth bears long, sharp, backward-curving teeth. The veil came from her body. The diamonds are air bubbles, and the pearls are her eggs. It is such a delicate and pristine object, more beautiful than any wedding dress train, yet issued from such an unlikely creature. It reminds me of the angelic voice of Susan Boyle, the dowdy, unemployed forty-seven-year-old who first appeared on the stage of *Britain's Got Talent* in 2009, and wowed the world with her singing.

Bill has known this goosefish for nine years, and he knew she was pregnant. This past weekend, he had led a group of teens on a camping expedition. The weekend was long and challenging. Nonetheless, he came in Sunday night to check on the pregnant goosefish. "I was getting nervous," he tells us, "because she was just massive." The goosefish before this one had grown to twice this one's size and had to be shipped to a larger tank in Quebec. Birthing her egg veil had caused a prolapse, and she'd needed surgery to repair it. The next year, her egg veil had gotten stuck inside her, like a human baby in breech position, and the vet performed a second surgery to remove it. The next year, she produced yet a third egg veil, and the vet removed both her ovaries. That tough old goosefish survived all three surgeries, but Bill wanted to spare this younger one such an ordeal.

"She was very uncomfortable," Bill said. "She looked like she'd swallowed a basketball. She couldn't even rest on the bottom." He was relieved when he came in the previous night to check on her again—and found her eggs floating in the dark water like the Milky Way in the night sky. Like Octavia's, these eggs are also infertile. But that does not diminish their unlikely provenance or their breathtaking beauty.

Everywhere, impossible changes unfold before our eyes. In the leafy sea dragon tank, the male gives birth to the babies, who come shooting out of a belly pouch that is like an opossum's. Among the corals of the Giant Ocean Tank, a species of fish called the bird wrasse starts out life as a black or brown female, then turns into a male. The commonest of sea creatures are miracles. Take the jellyfish. Many are born here; from eggs and sperm, they begin life as plankton, then turn into brown blobs and settle on rocks or docks, as polyps. They start out looking like something you'd scrape off the bottom of your shoe, then grow into something more beautiful than an angel.

"In the ocean, it seems that anything is possible," I comment one day as Christa and Anna and I stand watching the rays and turtles sweep past us in the GOT.

"Wouldn't you love to actually be in there with them?" says Christa.

"Wouldn't you love to be actually in the *real ocean* with them?" says Anna.

"Then, let's!" I suggest. "This summer—let's learn to scuba together!"

We announce our plan to Scott and Wilson over lunch at one of our favorite places, a hybrid Mexican-Irish restaurant named Jose McIntyre's. Scott thinks it's a great idea. Scott's job has included many scuba research and collecting expeditions. One of them was in the West Indies. Standard scuba safety procedure forbids a person dive alone, and Scott had promised to buddy up with a researcher whose quarry was active before dawn. "But almost everyone was out partying late at night," Scott tells us. Scott's partner was an excellent diver and didn't need monitoring; but still, Scott faithfully accompanied him to the dive site at four thirty each morning. They worked beneath the arch of an underwater cave in eight feet of water. Scott, exhausted, would strap on his tank, put his regulator in his mouth, inflate his buoyancy control vest, park himself beneath the arch among some brain coral—and sleep for two hours. Then his buddy would wake him up and they'd go back to the hotel. "But at night," Scott says, "I'd wake up in my bed and forget where I was, and thrash around trying to find my regulator."

I ask what to do if you need to cough or sneeze while diving. "No problem. They even teach you how to vomit correctly underwater," Scott replies. He says this actually happened when some party-ing visitors paid for a special pass to dive in the Giant Ocean Tank.

"Oh, my," says Christa. "What . . . ?"

"Tacos from this place," Scott answers.

✣

One of Octavia's arms is under her body. Another is attached by twenty-eight large suckers, some of them more than an inch across, to the rock ceiling of her lair. Another arm adheres by its suckers to the wall. The skin between Octavia's arms hangs like drapery. Then, at 8:25 a.m., her arms begin vigorously sweeping the skein of eggs farthest away from me, an athletic action that reminds me of a woman vacuuming blinds or curtains. She continues to work at this for two minutes. Then she turns and blows water at them with her funnel. No wonder the eggs are still so white. How does she avoid tearing the egg chains from their holdfasts?

The soft light goes on in her exhibit. The staff is gearing up for the public. Octavia's body grows as she draws water into her gills, her mantle expanding like a pink lady slipper orchid in bloom. I count the seconds between her breaths. Sixteen. Seventeen. Fifteen. One of the tendrils of her arms has turned into a knot. Then she unknots it to a corkscrew with three loops, as carelessly as a person would draw a doodle.

One huge breath takes three seconds to inhale and Octavia's whole body balloons. Only one arm is moving, her left front arm, cleaning the eggs at the back of her exhibit again.

At 9:10 I hear the first shrieking toddler of the day. My golden time alone with Octavia is about to expire. But the next hour at Octavia's tank is valuable for another reason: Though my view of Octavia is often blocked by jostling children and pushy adults, I can immerse myself in the tsunami of emotions, memories, and misconceptions that an octopus draws from visitors.

"There's the octopus!" cries a young woman.

"Beautiful!" says her bearded companion.

"Creepy, but beautiful!" adds a tall woman in back of the couple.

"Is that the octopus?" asks a little boy, pointing to the bottom of Octavia's tank.

"No, it's an anemone," his father replies.

"That's the octopus's enemy?" the child asks, worried.

I point out Octavia in her corner, and show the boy her eggs. "Wow!" he says, and then announces, "I'm a scientist, an animal rescuer, and an ocean explorer!" And with this, he runs off, parents in pursuit, to save the sea.

At 9:20, a family of three surrounds me. "Oooh! Octopus!" the mother says, reading the plaque by the tank. But they don't see Octavia at all until I point to her, and then I show them her eggs. They are tremendously excited. "Will there be babies?" the son, who looks to be about eight, wants to know. No, I explain—there was no dad, so there won't be babies. "Just eggs, like a hen lays eggs even when there's no rooster."

This distresses the boy. "She needs a dude!" he cries. His dad agrees. "Can't they fly in a dude octopus to help her out?" he suggests. A romantic thought, I concede, but, I tell the family, because octopuses may eat each other, blind dates in the confines of an aquarium, where nobody can jet away if they don't get along, are even more risky in captivity than in the wild.

"Can't they just inject the eggs with sperm?" asks the mother.

Unlike fish, octopus eggs must be fertilized before they're laid. And then, I note, there's the issue of what to do if the eggs did hatch: "What would you do with 100,000 baby octopuses?"

"Sell them to other aquariums!" says the dad, clearly an entrepreneurial type.

Everyone in the family seems so eager, almost desperate, for Octavia's eggs to hatch. In their refrigerator at home, there is surely a carton of unhatched chicken eggs, which produce no such distress. That's because the mother hen is nowhere in evidence. But this would-be mother is before them. There's a sweetness to these people's wish for Octavia's eggs. They seem happy in their family. No wonder they want Octavia to be happy too.

The arms farthest from us fluff her eggs. She frills her suckers, cleaning them, and flips upside down. Then she flips back. Two "horns"—actually soft papillae—rise above her eyes.

"Ew! I bet it feels gross if you touch it!" says a teenage girl. She's one in a pack of three, all in tight jeans and short jackets and thick eye makeup. I turn to face the speaker. Her young face is contorted with revulsion. "But look," I say, "did you see her eggs?" I point to the forest of tiny white globes hanging from the ceiling of her lair. "They're all eggs. There are thousands of them! And she is taking such good care of them all."

"No way!" says the girl who spoke first. "Way cool!" says one of her friends. Their expressions soften. Mouths that were twisted in disgust now open slightly; the pupils of their eyes dilate. "Yes—see how she's fluffing the eggs with her arms? That's helping to keep them clean and oxygenated."

"Aawww!" the girls now coo as if they are watching a puppy. A minute ago, Octavia was a slimy monster. Now that she is a mother, she's adorable.

"When will the eggs hatch?" the girls want to know.

I shake my head and explain that the eggs aren't fertile. In the eyes of one of the girls, I see a sheen of gathering tears.

I share a few facts about octopuses that I hope will interest and

impress them. I tell them about Octavia's venom, her beak, her camouflage, but the girls grow silent, their young faces stony. I'm losing
them.

And then Octavia inserts the tip of one arm into her mantle
opening. "Maybe she has an itch," I say. The girls' expressions grow
soft again. "Yeah," one says, and they share a sweet laugh.

They don't want to hear how Octavia is different from us. They
want to know how we're the same. They know what it's like to have
an itch. They can imagine what it's like to be a mother. This brief encounter has changed them. Now they can identify with an octopus.

They all take pictures on their cell phones. They thank me before they leave. "Take care of that little momma," one says to me
gently.

❧

As August arrives, it's time to get serious about scuba, before New
England's waters get too cold or too rough. Until Anna's fainting
spells are under control, scuba will be too dangerous for her. That
leaves Christa and me. I visit the dive shop Scott recommended,
United Divers in nearby Somerville, to register for the class, and
return to the aquarium at 6:15 p.m., to discover Teen Appreciation
Night in progress—a party the aquarium throws each year to honor
its young volunteers. I slip through the knots of chatting teens and
parents to Octavia's tank. Octavia is puffed up, her skin not, as usual,
creased or crinkled, thorny or warty, but smooth as a blown-up balloon.

This looks decidedly wrong to me, like a giant tumor or an internal organ bloated with disease. My distress increases when I can't
see her gills, her funnel, or her eyes. She has turned her face to the
wall, as a dog or cat often does when suffering. Except for part of
one arm, hanging down, all of Octavia's suckers are facing inward,

too, attached to her eggs or to the walls of her den. Her body color, in the red light of my headlamp, seems pale pink, veined with maroon, like spider veins on an old woman's legs. The webbing between her arms looks gray.

I'm beside myself with dismay. I've never seen her like this. Is she dying? There's no one I can contact, because there is nothing anyone can do. Female octopuses die within a few months after they lay eggs. No one can stop it.

But I don't want to see my friend die.

Suddenly Wilson is beside me, like an answer to a prayer. His granddaughter, Sophie, is one of the teens the evening program honors tonight. He didn't know I'd be here. He wanted to check up on Octavia.

"That is very weird," he says, looking at the octopus with concern. "I've not seen that texture before. But remember, you are seeing the end. If this is it, what are you going to do?"

I don't want to burden Wilson with my distress. After all, he is facing the same issue with his wife, whose situation is tragic as well as mysterious.

Wilson and I stand and watch the octopus, silent. Is Octavia thinking anything at all, and if so, could I understand her thoughts? What is going on in the separate, holy, mysterious, private theater of these minds? Can we ever know the inner experiences of another?

Learning, attention, memory, perception—these are all measurable, relatively accessible, amenable to study. But consciousness, says Australian philosopher David Chalmers, is "the hard problem," precisely because it is so private to each inner self. Other philosophers suggest that the self is an idea without basis. "Science does not need an inner self," writes psychologist Susan Blackmore, "but most people are quite sure we are one."

"The self," Blackmore writes, "is just a fleeting impression that arises with each experience and fades away again. . . . There is no inner self," she argues, "only multiple parallel processes that give rise to a benign inner delusion—a useful fiction." She argues that consciousness itself is a fiction.

The Buddha denied the existence of persisting selves. At the end of life, the self may dissolve into eternity like salt in the ocean. To some, this might seem distressing. But to lose the lonely self in the ocean of eternity could also be a release, an enlightenment, as the mystics promise.

⁂

At 7:05 p.m., one of Octavia's arms starts to move, slowly stroking the eggs nearest the window. She is still bloated, face against the wall, and we can't see her breathing. One of her arms is attached to the roof of her den with a single sucker, like a mosquito net hanging from one nail.

At 7:25, she has raised some papillae on her body. But they are few and low. Her skin is still smoother than either Wilson or I have ever seen it.

Then, at 7:40, Octavia suddenly twists around. I can see one eye, the pupil a slit. Wilson and I suck in our breath. Tall warts arise on her body and head. She inserts one arm inside her gill opening. Her arms begin to wave with great vigor; she turns and faces us. And in turning, she reveals her eggs to us—there are thousands of them!

Octavia seems to have snapped out of her stupor. Suddenly she spins her arms around and around, her white suckers swirling like the frilly petticoat of a can-can dancer. Forcefully she shoots water from her siphon, a typhonic sneeze. Out comes all sorts of whitish fibrous material. What is it? Excrement? Gunk stuck in her gills? And now Octavia resumes actively cleaning her eggs, stroking them with her suckers.

The crisis past, Wilson leaves to join his granddaughter. At 8:15, Octavia has spread her webbing like a blanket over the eggs and is hanging upside down, looking like a perfectly healthy mother octopus. Only a few of her eggs are visible now, looking like a necklace of tiny seed pearls on black string. At 8:20, she seems settled down to sleep. And soon I will, too. This night, I will stay in the hotel down the street. I want to see her the moment the aquarium opens its doors to the staff in the morning.

When I return, at seven the next morning, she is super thorny. She does not look anything like she did last night. She has remade herself. Mottled with dark patches, she is radiantly beautiful, the very picture of a healthy octopus and a diligent mother. She fluffs the clusters of eggs nearest the window with one arm, like a mom sitting on a park bench might jiggle a baby buggy. Who knows what she is doing with the other arms and other eggs I can't see? The lights haven't yet been turned on in her exhibit; without my headlamp, I would not be able to see her at all.

"I get so nervous every morning when I come in," says a voice beside me. It's one of the interns I haven't met before. "I'm so afraid I'll come in and find her dead on the bottom." Once a week, she cleans the octopus tank for Bill first thing in the morning, and she's noticed the eggs are shrinking. Some have fallen to the gravel floor of Octavia's exhibit. We don't know if Octavia has noticed this, or if this bothers her.

Everyone who works here knows the bittersweet news about Octavia's eggs. All the staff and volunteers now look upon Octavia with exceptional tenderness.

"Do you think she knows us?" asks the cleaning lady who wipes the glass of her exhibit each morning. "Does she know we're here?"

"Yes, I think so," I reply, "but I don't know how much she cares, now she has her eggs. What do you think?"

"I think she notices us. I know they're very intelligent," the woman says. "I notice her every day, and I think she notices me back. I can't say why."

Octavia's one visible eye, coppery now, not silver, faces us. I can't tell if she is looking at us or staring into space like a person lost in thought. She looks healthy and strong, but seems in a sort of suspended animation. Her breaths come at twenty seconds, twenty-four seconds, fifteen, eighteen. Is Octavia in an "egg zone," in which little else registers, like some young mothers? So many of my friends, once outgoing and social, are transformed once their babies are born. Women who couldn't sit through a two-hour concert are held transfixed by their infants, even though the babies do little more than suck, sleep, and cry. Hormonal changes that occur at birth, including a flood of oxytocin, popularly known as "the cuddle hormone," help make possible this change. Similar hormones might inspire Octavia's devotion. In fact, octopuses have a hormone so like oxytocin that scientists named it cephalotocin.

"All the hormones we've looked for in octopuses, we've found," Jennifer had told me when we'd met in Seattle. A paper presented at the Octopus Symposium detailed how researchers at the Seattle Aquarium found the hormones estrogen and progesterone in their female octopuses, testosterone in their males, and the stress hormone, corticosterone, in both. A female octopus's estrogen level spikes when she is of egg-laying age and meets a male. The male's testosterone levels rise.

Hormones and neurotransmitters, the chemicals associated with human desire, fear, love, joy, and sadness, "are highly conserved across taxa," Jennifer said. This means that whether you're a person or a monkey, a bird or a turtle, an octopus or a clam, the physiological changes that accompany our deepest-felt emotions appear to be the same. Even a brainless scallop's little heart beats faster when the mollusk is approached by a predator, just like yours or mine would do were we to be accosted by a mugger.

"Ew! An octopus—disgusting!" a boy about six years old cries behind me. And then another voice at my side: "She looks exceptionally beautiful today to me." It's Anna.

"I used to get here wicked early and spend all my time in front of her tank," Anna says, as the boy moves away. "That was after my best friend killed herself."

"Oh, Anna," I whisper. "How awful."

"She was so successful. She had so many friends. Everyone would have said if any of us would try to kill themselves, it would be me, not her," Anna continues.

Anna often feels uncomfortable in her body. She gets excruciating migraines, preceded by awful feelings that caterpillars are crawling up her neck. She has difficulty sleeping. She often has trouble focusing her mind and feels stupid. These are not uncommon problems for people on the autism spectrum. Add to that the hormonal turbulence of puberty, and the feelings can seem unbearable.

As we stand watching the octopus, she confides that, even before her friend had killed herself, Anna had tried to kill herself, too.

Shocked, I turn to face Anna as I gesture to Octavia. "You would leave *this*?"

"I wasn't involved with this place then," Anna says. "I wish I had known then that only five percent of the ocean has been explored. ..."

Her voice trails off, but I know what she is thinking. If only she could have conveyed the importance of this fact to her friend, perhaps everything would be different. For who would want to leave this vast, teeming blue world? Surely its waters could wash away all sorrows, heal all brokenness, restore all souls.

And in a way, for Anna, it has. Later, in an e-mail she wrote me at 2:30 a.m., Anna would tell me more.

"My best friend's name was Shaira," she wrote, "and I had seen her the night before." But the next morning, Anna was worried; Shaira had told her she was going to spend the night at her boyfriend's house; told Anna's parents she was going home; and told her own parents she was staying over at Anna's. But Shaira had slept at none of these places. And in the morning, Shaira didn't come home.

Periodically that Monday, people would call Anna with updates. Anna was feeding the Amazon catfish, Monty, when Shaira's sister called her cell phone to confirm that Shaira had not gone to her boyfriend's house. "The arowana bit me right then," Anna wrote, "but the catfish let me pet him. I was crying."

Anna was too upset to work, so her mother picked her up at the aquarium. In the car, Shaira's sister called to say she had found the suicide note. Shaira's body was discovered in a small pond ten minutes' walk from Anna's house, where she had gone to drown.

Anna called Scott and Dave to tell them she wouldn't be volunteering the next day, but they both said the same thing: She should come in if she felt like it might help. So she did. And she came in the day after that, too. She was working that Wednesday in Cold Marine when Dave suggested she might want to play with Octavia. "At that point," Anna wrote me, "I had already taken her out more times than I could count, and I felt like I knew her pretty well. I think she sensed something was wrong. She was a lot gentler than she usually was, and she had her tentacles on my shoulders. It's hard to explain why I think she understood. . . . After interacting with an animal lots of times, you get to understand what the usual behavior is and what it does in different situations.

"I find myself expressing my feelings more when I'm around her," Anna wrote me. "When I'm sad, my tremor gets worse. My arms get weaker, and my body temperature drops. When she came out, I felt like I could stop holding my breath. I cried, but then I stopped crying, because there was an octopus on me."

Except for the day of Shaira's funeral, Anna never missed a single day volunteering at the aquarium that week, and in fact earned Volunteer of the Month that May.

The rest of the school year was difficult; occasionally Anna tried to escape from her pain by using drugs. But she would never use drugs at the aquarium; she wouldn't even think of using drugs the day before coming in to volunteer. "My logic was that, at the aquarium, I didn't want to be anything but there," she said.

"This has been the worst summer of my life," Anna wrote me, "but my days at the aquarium have been the best days of my life. I've learned," she said, expressing a wisdom way beyond her years, "that happiness and sadness are not mutually exclusive."

This names the way we feel as we watch Octavia, our alien, invertebrate friend, caring for her infertile eggs at the end of her life with a tenacity and tenderness at once heartbreaking and glorious.

In *The Secret Garden*, Frances Hodgson Burnett writes of the beauty and solemnity of eggs: "If there had been one person in that garden who had not known through all his or her innermost being that if an Egg were taken away or hurt the whole world would whirl round and crash through space and come to an end . . . there could have been no happiness, even in that golden springtime air." Eggs were surely life's first love, and protecting one's eggs was surely love's first urge. Love is that ancient, that pure, that lasting. It has persisted through billions of species, through millions of years. No wonder the sages say that love never dies.

And Anna well knows this truth. As Octavia tends her infertile eggs, Anna tends her young friend's grave. She looks for special, beautiful rocks to bring to the cemetery, she tells me. She knows that love lasts through everything, and that not even death can erase it.

Though Octavia's eggs will never hatch, it fills us with gratitude that Octavia tends them with diligence and grace. For when she dies, Octavia will do so in the act of loving as only a mature female octopus, at the end of her short, strange life, can love.

❊

By the end of August, Octavia is still active and strong. Bill tells me that, as he was feeding the anemones and the sea star in her tank, she had reached out her arm and hungrily grabbed two capelin right out of their tentacles and eaten them the day before. "She could last a long time," Bill says, "and that's one reason I want something else for Kali."

Kali has been, as Wilson puts it, "acting up." First she was squirting. Next she bit Anna. Then she started demanding fish from us at jet-point. Lately, she has been acting oddly. When we remove the top of her barrel, she ascends to the surface but won't linger. She sinks back down, turns pale, and watches us from the bottom. I ask Bill if he's worried about this. "Not yet," he says.

This is exactly what she does when Wilson and I visit her. She balloons to the top, her mantle inflated, the webbing between her arms billowing like sails. But she does not show us her underside, asking for a capelin. Wilson flips over one of her arms—her second left arm, known as L2—so the suckers are up. He hands her a fish, and she seems to accept it. But rather than let us watch her eat it, like she used to, she descends. She drops the fish. And strangely, she seems to want to look at us but not interact with us. Wilson closes the lid.

When Wilson and Christa and I visit her that afternoon, she's at the top of the tank, waiting for us when the lid comes off. She sucks on our hands gently for half a minute. She pauses when her suckers touch the Band-Aid on my thumb, recognizing this as something new, touching it tentatively. I wonder what the adhesive tastes like to her. She soon lets go of us. As she drops to the bottom, my heart sinks with her. Is she sick? Has she already seen too many people? Is she despairing in her small, bare barrel? Doesn't she care about us anymore?

But then when I step away from the tank to speak with Bill, Kali rises up instantly and turns bright red. Is she looking for me? Wilson calls me back. I stroke her head and she stays with me for many minutes before dropping again to the floor of her barrel. She stares up at us, her eyes inscrutable.

Wilson is worried. After I leave for the day, he speaks with Bill.

"The number of contacts Kali has is significantly higher than any other octopus. Are you with me, Bill?" he asks.

"Definitely."

"Last week, everyone was there before us. And I say too many other people are coming."

Bill agrees. He's been thinking about this very problem. Research including Jennifer's shows that wild octopuses choose to spend 70 to 90 percent of their time crammed into tight dens. But that still leaves time for Kali to be bored. When people want to interact with her, if she's not in the mood, she can't get away. She can't hide in a lair like Octavia can in her big tank. In the barrel, Kali is, as Wilson puts it, "a sitting duck."

A few weeks ago, with Bill's permission, I gave Kali a clean terra-cotta pot in case she wanted to hide in it. At the Middlebury octopus lab, the octopuses so valued these pots they could be used as a reward for correctly negotiating a maze; but to our knowledge, Kali has never used hers. We've never seen her hiding in it. She's always at the top of the barrel when we open the lid. So Bill removed it. The pot seemed to be just taking up space—and space in the barrel is getting short. Kali is two thirds the size of Octavia now.

Bill's options are limited. He can't put her in the big tank with Octavia. One octopus would almost surely try to kill the other. "I want her in a new system," Bill tells Wilson.

Karma unfurls herself like a silk scarf in the water of her exhibit. Her red color shows her excitement. (© TIANNE STROMBECK)

Though Karma was always gentle with us, her 1,600 suckers were capable of tremendous strength. One scientist calculated that the suckers of the much smaller common octopus could exert a quarter ton of force. (© TIANNE STROMBECK)

Octavia reaches out of her exhibit to embrace Anna. (PHOTO COURTESY MAGILL-DOHAN FAMILY)

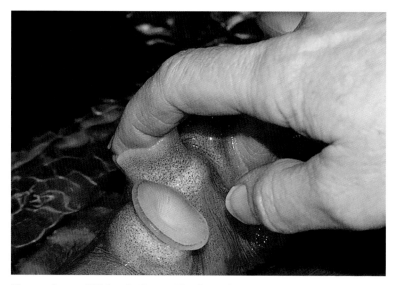

Karma kisses Wilson's finger. Each sucker is capable of an exquisite sense of taste as well as a pincer grip so fine the animal can untie knots. (© TIANNE STROMBECK)

Myrtle the green sea turtle is the undisputed queen of the Giant Ocean Tank. Even the sharks (including the bonnethead shark, behind) defer to her. (© JOHANNA BLASI)

Smallmouth grunts school in the renovated Giant Ocean Tank. (© JOHANNA BLASI)

An octopus's mouth is in its armpits. Octopuses generally grab prey with their suckers, then pass it from sucker to sucker, as if along a conveyor belt, until it reaches the mouth. Here, Karma enjoys a fish.

(© TIANNE STROMBECK)

Karma, white and calm, at home in her 560-gallon exhibit. To her right is the male sunflower sea star, who many mistake for an octopus. (© TIANNE STROMBECK)

A southern stingray glides by like a magic carpet. Flattened relatives of sharks, several species of rays share the Giant Ocean Tank with the gray angelfish, yellowtail snapper, and parrotfish seen here, and more than 100 other species. (© JOHANNA BLASI)

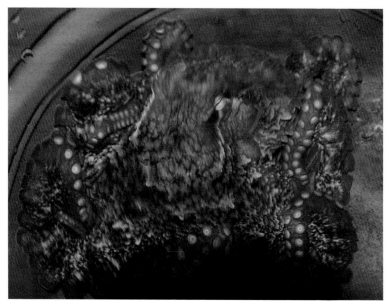

Kali, aptly named after the Hindu goddess of creative destruction, looks up at us mischievously from her barrel. (© TIANNE STROMBECK)

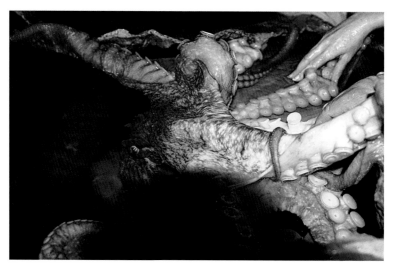

Wilson and I commune with Karma. (© TIANNE STROMBECK)

Octopuses are curious about their neighbors. In Mooréa, Keith saw an octopus eyeing this grouper with evident interest.
(© KEITH ELLENBOGEN)

In Mooréa, one octopus (left) moved into position for a better look at Keith as he took photographs. Note the second octopus, at bottom right. (© KEITH ELLENBOGEN)

The author swimming with a wild octopus in Mooréa. Though the female in the foreground had lost parts of several arms to a predator, she was still wonderfully curious and bold. (© DAVID SCHEEL)

Octavia cradles her eggs, the light-colored masses hanging like clusters of grapes and spilling out from between her arms in the upper left of center. What some visitors may mistake for two closed eyelids is actually the horizontal pupil. (© TIANNE STROMBECK.)

But that's easier said than done. "All the dominoes have to fall into place," as Wilson puts it when he tells me later. "You want to move this fish, but you have to move this other fish first, and before that you have to move this other fish, and because they're moving, these other animals over there have to move," he said. "What you get and when you get it is something you can't always control." Every day, animals at the aquarium are being born and dying, arriving from collection expeditions or from U.S. Fish & Wildlife Service agents, or getting shipped to and from other aquariums throughout the United States and Canada.

The comings and goings are always delicate, frequently surprising events. One morning I find Bill has been gifted with a 21-pound lobster caught off Nauset Beach in Orleans, Massachusetts—given by the anonymous winner of a raffle at Cap'n Elmer's fish market to benefit Dana-Farber Cancer Institute. The lobster's claws are so heavy he cannot lift them out of water. Another day, eighteen Amazon stingrays arrive in Freshwater, each as large as a bathmat. They had been living in a huge tank owned by a paraplegic man whose ground-floor apartment is being renovated; they have grown too large for him to keep. (He was grateful the aquarium could take them, but wept when the aquarium van pulled away.)

And then one Wednesday I come upstairs after watching Octavia to find Scott's team catching angels.

There are twenty-six of them. As well as sixteen plecostomus, one lilith, seventeen Geophagus, two silver arowanas, and a host of other species. The angelfish and their tank mates are being moved, after a year of planning, from behind the scenes, where they have thrived since their arrival from the Amazon, to the Amazon exhibit out front. The big round tank out back, behind the scenes, has been largely drained so that Christa and another volunteer, Colin Marshall, both in wet suits, can catch the fish in shin-deep water with their nets, working by chasing the fish toward each other. Colin and Christa transfer each fish in the net to Scott, who's wearing his wet suit too and holding a net of his own. As he transfers each to the new tank, he calls out the species name, and since there are often several fish in each net load, how many there are. Anna records their number as Wilson, Brendan, and I watch and try not to get in the way.

It takes an hour to move the fish. Immediately we troop down to the public area to see them in their new home. I have never seen Scott so tense. He was up all night last night, worrying. "Fish could be eaten. Fish could die of stress," he says. But the moment we arrive in front of the tank, Scott falls silent. "He's watching the fish language," Wilson whispers to me. The angels' stripes are lighter than usual, a sign of distress. But happily, an hour later, the normal dark color is restored. They're even eating in the new exhibit. Scott breathes a sigh of relief.

Another Wednesday, I arrive to find Bill has just rearranged all the rocks and moved the purple sea urchins, giant acorn barnacles, whelks, giant green anemones, feather duster worms, and tube anemones in the Pacific Northwest exhibit next to Octavia's. He's pleased with the rearrangement, which looks great—but he hates to upset his animals. "They've been here longer than me," Bill says. His purple sea urchins can live about thirty years; the feather duster worms survive a century; and the anemones, if unmolested by predators or stricken with disease, can theoretically live almost forever; scientists note that they do not appear to show signs of aging.

But these potentially long-lived animals can be finicky, especially the delicate anemones. When conditions are right, each opens like a beautiful flower in bloom, capturing nutrients with its petal-like tentacles. When disturbed, it withdraws into a compact blob nobody would notice. These animals have no brains and the barest of nervous systems. And yet their behavior is eloquently expressive. Neurosci-

Check Out Receipt

Saline District Library
734-429-5450
http://salinelibrary.org

Wednesday, April 26, 2023 7:48:29 PM

Title: The soul of an octopus : a surprising exploration into the wonder of consciousness
Call no.: Book Disc. 594.56 Mon c.14 Great Summer Read
Due: 5/24/2023

Total items: 1

****
You can now pay your library fines
online at https://www.salinelibrary.org

All overdue MeL fines are 50 cents/day

For self-serve hold questions call the
Circulation desk at 734.429.5450

entist Antonio Damasio briefly mentions anemones in his book on consciousness and emotion, *The Feeling of What Happens.* He does not argue that anemones possess consciousness; but, he writes, in their simple, brainless behaviors, we can see "the essence of joy and sadness, of approach and avoidance, of vulnerability and safety."

"The anemones might not like where I put them," Bill says worriedly. One of the four tube anemones, whose tentacles were still withdrawn yesterday, has opened up today; but a white spotted rose anemone is still unhappy and hasn't opened. "Everyone you disturb takes a while to recover," Bill explains.

But the most disturbing disruption in the history of the aquarium lies ahead. The Giant Ocean Tank, the centerpiece of the aquarium, is going to be rebuilt from the top down. Four hundred and fifty animals of one hundred species will be moved—more than half the animals in the aquarium. Space that's already tight will be at a premium. For the next nine months, nothing will be routine. It's going to make finding space for a large, new, octopus-proof tank for Kali fantastically complicated.

❧

"This is the biggest project the aquarium has ever done since it was built," Billy Spitzer, vice president of programs and exhibits, announces to the staff and volunteers gathered for the brown-bag-lunch presentation on this last Wednesday in August. He's wearing a hard hat and orange safety vest for effect. "This is even bigger than when the aquarium was built—because it's being done while it's open to the public."

Many of the animals to be moved—including Myrtle and her chelonian colleagues, the sharks, rays, and moray eels, as well as hundreds of large and small reef fish—will be relocated to the shallow penguin tray. Its 110 gallons, chilled to 61°F for the penguins, will be heated to 77°F for the tropical fish. The eighty African and rockhopper penguins were already moved last week to the aquarium's Animal Care Center in Quincy to make way. The little blues will move to a corner in the New Balance Foundation Marine Mammal Center on the first floor. The whale skeleton is coming down so the ceiling can be fitted with new lighting. After more than forty years of salt water and pressure, the sixty-seven glass panels of the Giant Ocean Tank spiral will be removed and replaced by clearer-than-glass acrylic panes. Over the course of the next nine months to a year, two thirds of the two thousand sculpted corals in the tank will be removed and replaced with two thousand new, softer, more colorful and easier-to-clean coral sculptures. And when the $16 million project is finally done, the GOT will be renewed from top to bottom. It will be more accessible and offer better visibility. And with many new additional hiding places for fish among the new coral sculptures, the tank will host nearly twice as many animals.

"It's a great opportunity," the vice president told the staff, "but we've got to admit, it's stressful." Some staffers are already mourning the changes. Some speak of "the psychology of loss." For nine months, no penguins will greet the staff or public as they enter the building. For much of the year ahead, the aquarium will be bereft of its centerpiece. Some of staffers' favorite animals will be moved off-site. People and animals will be cramped to make room for construction workers in hard hats and their tons of equipment. What was once beautiful will become ugly. What was once tranquil will become noisy. Nothing will be the same.

Next Tuesday, he tells us, the aquarium's transformation will begin.

And although I didn't realize it then, a transformation of my own would soon follow.

CHAPTER FIVE

Transformation

The Art of Breathing in the Ocean

I am drowning.

Well, not quite. But there is water in my airway, I am 14 feet underwater, and it appears even more water is pouring in. My usual response to such a state of affairs, one that has worked well for over half a century, is to get my head out of the water and gasp. But my scuba instructor is horrified.

"No, no, no! You must not surface so fast!" the young guy with a French accent admonishes once I bob back to the life-giving air.

"I'm sorry," I gurgle. "Water was coming into my regulator. Why is this happening?"

Later that day I learn from another instructor that my problem was the same that sinks ships: loose lips. I need to grip the regulator more strongly with my lower lip, which I am not doing very well because, apparently, I am spending a lot of time underwater actually *smiling*. Here in the MIT swimming pool, I am practically delirious, reveling in my amphibious transformation. I am imagining my not-so-distant future, swimming among corals, among fishes, among sharks and rays and moray eels and most of all, among octopuses. I can't help but grin like a lunatic.

But nothing wipes a smile off the face as fast as nearly drowning. The French fellow admonishes me, "Baby steps!" But for me, the very idea of scuba is a giant leap from everything I have previously known.

I am taking the intensive scuba course just outside Boston, as Scott had suggested—alas, without Christa, who had to cancel at the last moment. Though I miss my friend, I wasn't worried about taking the class. I have always been somewhat aquatic. I'm not a particularly graceful or strong swimmer, but I'm a fearless one. From the Gulf of Siam to the murky waters of the Amazon, I have always felt confident I'd be fine if I obeyed Rule One of swimming, and that is: Don't Try to Breathe Underwater.

Except now this is exactly what we are supposed to do.

Everything in scuba is completely different from life on land, and also from previous experience swimming. Scuba equipment is intimidating and heavy. Just assembling the gear—the nearly 40-pound air tank, the vestlike buoyancy compensation device, or BCD, with its additional pounds of lead weight in the pockets, the hoses and gauges and mouthpieces that hang from tubes everywhere like sleepy eels—takes seven complicated steps to accomplish. Screw up any of these and you have a big problem. Yet the assembly still seemed to me—a person who had managed to make it through not just one, but two high schools without mastering the combination to my locker—an impenetrable mystery.

In my rented gear, my body feels utterly alien. The giant fins are big as clown shoes, the mask prevents peripheral vision, and breathing through the regulator in my mouth makes me sound like Darth Vader. The BCD has air pockets that you can inflate or deflate to make you float or sink in ways you never did before. I'm wearing a rented mask that somebody else spit in (you spit in your mask to defog it), a wet suit somebody else peed in (we're told everyone does—in the ocean, not the pool), and a regulator that somebody else may have thrown up in. And in this gear, I'm not even supposed to swim normally. I'm supposed to fold my arms in like a kangaroo and propel myself only by kicking my fins.

Nothing looks right: Objects seem closer and 25 percent bigger in the water. Nothing sounds right: Sound travels four times faster in water than in air, and directionality is distorted. Nothing feels right: Since you're not really swimming, you can't warm up, and water carries heat away from the body twenty-five times faster than in air. Even though the pool is 80°F and we are wearing wet suits, everyone's lips are blue with cold by the end of the first session.

Yet I was doing the impossible and having a blast.

It was only when the water started filling the regulator that I panicked.

I was certain things would get easier as the weekend progressed. But I was wrong.

⚜

Everyone was spent after the first day of scuba lessons. Even our main instructor, fit, twentyish Janine Woodbury, admitted she was pretty beat. And, she confessed, her ears hurt. Mine too. Bad enough that I had taken a sleeping pill the night before, because the pain was keeping me awake. (I later learned this was dangerous and could have damaged my heart or lungs.) But on hearing (though more faintly than usual) my young instructor's admission that her ears hurt too, I felt better. Maybe your ears are *supposed* to hurt. In this I was wrong.

The pain did not decrease, but to my surprise, I was able to assemble the gear. I didn't have to think twice to remember how to clear my regulator or inflate or deflate my BCD. I felt strong and ready to learn new skills, including breathing from a buddy's emergency air supply, which is, to my delight, called an octopus. But my ears felt as if they were going to explode.

Janine had once actually seen one of her students' eardrums explode. "Bubbles came out of his ear underwater," she said. "It was gross." And incredibly painful. Unfortunately, permanent ear damage from scuba is not as rare as one would hope. Scott no longer dives because of ear damage that he suffered on a relatively routine expedition in Massachusetts waters, collecting live rock—pieces of dead coral that have been colonized with algae and sponges that, back at the aquarium, work as biological filters in tanks—at a depth of 100 feet. While surfacing, he suffered a "reverse squeeze" from pressure changes, and damaged his cochlea so badly his doctor forbade him to dive again.

I made the gesture for "trouble clearing ears" underwater to my instructor. She gestured for me to hold my nose and blow—a trick for equalizing pressure called the Valsalva maneuver. I did so with great enthusiasm and heard a very loud noise from inside my head. "Okay?" she signaled. But now it hurt worse. I signaled "something's wrong" and pointed again to my ears and blew again.

I ascended a couple of feet and tried the Valsalva again. I also used the Frenzel maneuver, another way of opening the eustachian tubes, by moving the jaw around like a snake trying to swallow something bigger than its head. This didn't work either.

"Okay?" Janine signaled.

No, I replied with my hand. I Valsalvaed yet again. I tried sinking a little—maybe this was a "reverse squeeze" and could be solved in this manner. But no—if anything, this made it yet worse. I rose again, slowly, pinching my nose and blowing the whole time.

"Okay?"

But I was not okay. No matter what I did, the pressure in my ears was blindingly painful.

I got out, and sat with my eyes closed, doubled over. Pain was not the sole source of my suffering. It was the prospect of defeat. I desperately wanted to be able to enter Octavia and Kali's world. Encumbered with my awkward skeleton and air-hungry lungs, I could not begin to discover much of anything about what it might be like to be an octopus without at least learning how to breathe underwater. I wanted to meet octopuses who lived in the wild ocean. In the shower, I began to repeat in my head the first words of the Fisherman's Prayer, the words to which John F. Kennedy kept on his desk at the White House: "Oh God, Thy sea is so great and my boat is so small. . . ." I fervently longed to get out of that boat and enter the Creator's great ocean, if only for an hour at a time, as a breathing, swimming sea creature. How could I do this without scuba?

Then I noticed the vertigo and nausea. When Scott suffered his cochlear damage, the pain had been accompanied by dizziness. When he surfaced he had thrown up.

But I was determined to try again. My instructor suggested Afrin, a nasal spray pilots often use. I wobbled to a drugstore and bought some, along with a macrobiotic bento box for lunch. The lunch did not stay down.

Janine gently suggested I bail. I did not want to lose my hearing. I have three smart, resilient friends who are deaf, and even they have a hard time in the hearing world. So I agreed. I would go home early, having failed to complete the first half of the course.

Defeated, I crawled off to my car, where I discovered I was too dizzy to drive.

I lay down in the backseat, on the blanket where our border collie sits, her paws and belly often plastered with mud, on the drive after our hikes in the woods together. Inhaling her scent, I felt instantly calmer. Within half an hour, though my ears still hurt horribly, the vertigo had eased enough for me to drive the two hours home.

❦

When I returned to the aquarium the next Wednesday, everything had changed. The top floor of the Giant Ocean Tank was closed to the public, and the walkway along the big tank was now draped with white cloth to screen the work under way. Eighty-gallon plastic tubs were scattered around, ready to receive fish. The top floor was littered with large wooden boxes in which the larger section of the reef would be carted away.

Octavia was in a strange spot, way farther back in her lair than usual, and at least fifteen rows of egg chains were visible, some of them nine inches long. She hung from her arms as if lying in a hammock, and she was exceptionally still.

Everything, in fact, seemed unnaturally subdued. The aquarium was nearly deserted. Visitors were few. Scott was at a conference in Tucson; Bill was on vacation in Florida; Anna was back in school. The penguins were gone, and their tray held only Myrtle and her fellow turtles.

Myrtle had been moved only the day before. A diver lured her with lettuce toward a white, sea-turtle-size plastic crate, with floats at the handles and holes through which water could flow. As Myrtle chewed her treat, another diver simply spun the 550-pound reptile around by her shell and gently pushed her inside the crate. She was then hoisted out of the water, wheeled into the elevator, and lifted into the penguin tray. Myrtle's legs began to whir as soon as the water flooded into her crate, and with a tip of her box from one of the four divers attending her release, the calm old turtle swam out confidently to her new home, apparently unfazed.

Myrtle's transition had been more successful than my own. I had hoped to return from my scuba weekend triumphant, a changed creature. But when Christa and Wilson asked me, I was forced to confess I was a washout.

Wilson was sympathetic, having tried scuba himself once. "This is not an easy sport," he said. Both his daughter and son are skilled divers, having logged many dozens of dives—including one on which a fellow diver died of decompression sickness.

I shared details of my failure as we visited Kali. She was at the top of the barrel before the lid was off, her body a rich, reddish brown, staring at us with golden eyes. Unlike the previous week, she was quite energetic, her arms reaching and grabbing at us with her suckers. "Easy, honey!" said Wilson, hurrying to feed her a squid and two capelin. Her suckers conveyed the food to her mouth in seconds, and in a minute it was gone. Then she turned her attention to playing with us, grabbing and pulling. Each sucker hugged and kissed us at the same time. I felt consoled.

Christa, always cheerful, was upbeat about my scuba setback. "I know you'll be able to do this!" she assured me. And in fact, I had already set in motion my next plan. About halfway along my drive to the aquarium is a dive shop, Aquatic Specialties, in Merrimack, New Hampshire, and I had arranged for private lessons, starting next week—so it would be possible to complete my open-water certification before New England's waters get too cold or too rough. My instructor was a volunteer at the aquarium, which I took as a good omen.

In fact, at the aquarium, everyone who worked on Tuesdays knew my new instructor. They called her Big D. Doris Morrissette, fifty-nine, red-haired, and possessed of mischievous humor, is only five foot one, but her spirit is huge. And she's an exceptionally patient and effective instructor because, she cheerfully admits, if you can make a mistake, she's done it.

As a child, she'd been enchanted with Jacques Cousteau and *Sea Hunt* on TV. But even though she was a strong swimmer and loved the sea, it didn't occur to her until she was fifty that she might be able to scuba dive herself—because all the divers on TV were men.

Finally she took a "Try Scuba" mini-course on a Caribbean vacation. After about thirty minutes of classroom instruction, her group went out on a boat, suited up, and jumped overboard. "Except me," she said. "I wasn't even in the water and I freaked out. I just couldn't do it." She tried again, taking lessons, working with two personal instructors and a nutritionist to get strong—and the next year got certified.

By 2010, she was an instructor. Since then, Doris has taught many dozens of grateful students. She leads weekly summer dives in New England and dives around the world. By the time I met her, she'd completed 375 open-water dives and, since she started volunteering at the aquarium in 2009, 180 dives in the GOT.

My two sessions with her in Aquatic Specialties' smaller, shallower pool were easy and fun, but as autumn advanced, I grew increasingly anxious about completing the four open-water dives I needed to finish the course. Big D had been forced to abort the last two dives she'd planned in Atlantic waters due to strong surf. But for me, she had a solution: I would earn my open-water certification in New Hampshire's Dublin Lake—only minutes' drive from our house.

Unfortunately, it was by then October, and the spring-fed lake was 54°F.

The ancient Spartans believed that cold water is good for everything, including your hair. Water temperatures in this range do, in fact, cause physiological changes—one of which is known as the cold-shock response, a "series of reflexes that begin immediately upon sudden cooling of the skin following cold-water immersion." During this reflexive response, "blood pressure, heart rate, and the workload of the heart all increase, making the heart more susceptible to life-threatening rhythms and heart attack. Simultaneously," an online text explained, "gasping begins, followed by rapid and deep breathing. These reflexes can quickly lead to accidental inhalation of water and drowning. This rapid and seemingly uncontrollable over-breathing creates a sensation of suffocation and contributes to feelings of panic. It can also create dizziness, confusion, disorientation, and a decreased level of consciousness."

I'm glad I didn't know this at the time.

To keep from freezing in New England's chilly waters, the scuba diver dons quite a bit of neoprene: I was to rent a seven-millimeter-thick, overalls-type wet suit, over which I would wear another seven-millimeter, long-sleeved, short-panted "shortie" wet suit. Pulling the legs of the overalls on was as difficult as it was awkward, with much tugging and grunting, but Doris assured me it would be worth it, because the harder it was to get on, the better the fit—and the better the fit, the warmer I'd be. But since the shop did not have a huge selection of rentals, and women customers are rarer than men, I was issued a men's size small. Of particular note was the roomy crotch, which caused me to adopt the walk of a woman whose panty hose are falling toward her knees.

I would also need to buy boots, gloves, and a hood. Putting on the hood is like pulling a plastic surgical glove over your head. It bent my ears in half like a pita bread around a falafel, and I felt certain I would smother. The neck was so tight I felt like my head was going to pop. Once the hood was on, I hoped it would smooth and stretch the skin on my face, giving it the pleasing look of a face-lift; but instead it squished my cheeks toward my nose, as if my head had been caught between the closing doors of an elevator.

Another feature of extra neoprene is it increases buoyancy, so the diver needs to don more weight. So in addition to the more-than-30-pound tank and the weights I was already wearing to dive in the pool, now I would have to string even more lead on a belt around my waist. This brought the total amount of extra pounds I would carry to nearly 70 . . . 57 percent of my body weight.

The increased weight, the cold, the extra gear, and the low visibility of its turbid water make open-water scuba in New England practically a technical dive. Both Doris and my previous instructor, Janine, both said the same thing: "If you can dive in New England, you can dive almost anywhere."

Big D and I loaded the gear into our cars and drove the hour from Merrimack to Dublin. Again I thrashed my way into the two-piece men's outfit. Struggling into my neoprene wardrobe by the side of busy Route 101, a road often traveled by friends and neighbors, I prayed nobody I knew would be driving past and recognize me right at that moment.

Finally dressed, I thought, Okay, this is so uncomfortable I won't even notice the cold water. I staggered into the lake, stepping from rock to rock to muddy bottom, and for a moment, I was dry and warm. Then the water began to seep in. I remembered longingly that Janine had told us that there are only two kinds of divers: those who pee in their wet suits, and those who lie about it. Ninety-eight point six degrees would have felt mighty good about then. I wished I'd had more to drink before I came.

That first day was foggy and rainy, but Big D was chipper: "The raindrops look awesome when you look up from underwater," she said. I dived beneath the surface only to career between sinking to the bottom and shooting to the top. My legs cramped in the cold. In the murky water, if my instructor swam more than ten feet away from me, she would vanish from sight.

Miraculously, I was able to perform all the scuba skills to Doris's satisfaction. We surfaced after twenty minutes, and she announced our next dive would be "just for fun." We could look for the big bass with which the New Hampshire Fish and Game Department stocks the lake; there are also landlocked salmon there. In the murk, we saw none. But Big D was right: The raindrops *did* look awesome viewed from beneath.

For the final dive, two days later, I practically fell back into the water. This time I was not even looking for fish. I just wanted the next dive to be over.

But then, a six-inch bass swam *right up to my face mask*.

It was unlike any encounter I'd ever had with a wild animal. Normally, you first see the animal at a distance; if you're lucky, gradually it might approach you or allow you to come close. They don't just appear inches from your face and stare at you. The bass may have been surprised, too. Some say that because their faces aren't as mobile as ours, fish don't have expressions, but they are wrong. His look was quizzical: "What are *you* doing here?"

We held each other's gaze for many seconds. Then one of us blinked. Since I was the only one with eyelids, it must have been me. The bass vanished as suddenly as a shudder.

But the fish should have been happy, because that day—the day I earned my certification—I was the one who was hooked.

❊

When I return to the aquarium, I find the last fish have been evacuated from the GOT. Aquarium engineers pulled the plug on the 200,000-gallon tank at 10 a.m., October 2, draining it at the rate of one inch per minute. Finally, divers could use ladders to reach the lower water level and catch the speedy tarpon, permits, and jacks in nets. While I was diving in Dublin Lake, Bill was on the crew that worked from 3 to 9 p.m. over the weekend, moving the eight four-foot-long, 40-pound tarpon. "They're big. They're hard," he said. "That's why they left them till last."

Every move is fraught with drama and danger. In September, a team of four divers, three veterinary staffers, a thirteen-member bucket brigade, one curator, and a handful of volunteers worked together to move the two three-foot-long blacknose sharks, a male and a female, from the GOT.

For weeks, divers had been acclimating the sharks to the nets, holding the nets in the water so the animals would not fear them. The team had successfully moved the bonnethead sharks the day before; but the blacknose sharks are more sensitive, explained curator Dan Laughlin, and might freak out. A frightened shark is nearly impossible to catch, which is why Dan had briefed everyone not just on plan A but, in case that didn't work, plans B, C, and D. (Plans B and C involved crowding the sharks by cutting off their usual swimming areas with nets or partitions; D meant waiting till the tank was nearly drained.) Worse than frightening a shark is injuring one, and this can happen easily if the shark thrashes against the sharp edge of a coral sculpture. "Don't strike," Dan warned the two divers armed with big scoop nets, "unless you're sure you're going to get them."

The plan was simple: The two netters, one of whom I recognized as Myrtle's friend, Sherrie Floyd, another, Quincy husbandry aquarist Monika Schmuck, would face each other, standing on opposite coral sculptures, in the center of which was a deep trough. A third diver, hovering in the water of the trough, would tempt the sharks with a herring on a pole. Once the treat captured a shark's attention, the diver would swing the pole toward the familiar net—into which everyone hoped the shark would eagerly swim.

At first the sharks seemed uninterested in the herring. They made one pass, then two, around the pole. Then a third. But the team had made sure the sharks were hungry. At the fourth pass, the female blacknose swam right into Sherrie's net. Sherrie scooped it up in one fluid motion and handed it to another staffer on dry land, who ferried the shark to the tank—which had been filled with salt water by the bucket brigade, assisted by a single pump—already waiting in the elevator.

The second shark, everyone thought, would be harder to catch. But just two passes later, the male was in Monika's net. Because he was larger than the female, and strong enough to flop out, our hearts stopped until someone clapped a second net on top of the first to prevent his escape. The two sharks were on the truck to Quincy before the divers even got into the showers.

Alas, the tarpon transfer had not gone so easily. They'd had to dissolve anesthetic in the water to slow the fish. One tarpon didn't recover from the anesthesia and died.

This was hard on Bill. On an earlier visit, I had watched him hold tenderly onto one of his more elderly charges, a redfish, while vet techs fed him through a tube. "He's not been eating," Bill told me with deep concern. The redfish's problem was a common one: He had a gas bubble in the eye, and the pain had ruined his appetite. He was being treated with steroid eye drops for the bubble, but as he recovered, it was important he keep his strength up. Bill was visibly tense until the redfish recovered from the drug and could be returned to his tank behind the scenes, which he shared with a fellow redfish and a brown eel known as a rock gunnel, both species common in nearby Maine.

Animal-keeping institutions aren't all the same in the care they give sick inmates. When a friend of mine was working at a small zoo in the early '80s, their kangaroo fell ill. She called a zoo in Australia for help. "What do you do when your kangaroo gets sick?" she asked. "Shoot it and go catch another one," came the reply.

But at the New England Aquarium, each animal, no matter how common, receives compassionate, expert care. Everyone loves these animals; nobody wants to see them suffer or die. One of Bill's surf perches was recovering from an episiotomy. These fish give live birth and her babies, stuck inside her, had ruptured her cloaca, exposing her intestines. The aquarium's boyish, cheerful vet, Charlie Innis, had operated on her with the same focus and urgency he brings to saving the dozens of injured, critically endangered wild sea turtles the aquarium rescues, rehabilitates, and releases each year.

The four-inch surf perch took a month to recover from the surgery. Today, Bill gently scoops her from her recovery tank and places her in a blue bucket, where she'll be anesthetized so two gowned and gloved vet techs can remove the stitches. One holds her on a yellow sponge towel while the other snips the sutures. Soon she will be released to a tank behind the scenes, which she'll share with some sea pens, beautiful, soft coral animals whose feathery appendages look like old-fashioned quill pens. Bill shows me the tank. It's next to one that once held lumpfish, and after that ocean pout, and now is filled with white anemones—animals that had only recently been moved there from the Pacific Northwest exhibit. This is the tank to which Bill would like to move Kali.

But when? The little octopus isn't so little anymore. When we visit her, she is active and affectionate, her suction leaving red hickeys on our hands and arms, but we all worry that in her small, boring barrel—with nothing to play with, nowhere to hide, nothing to see—she could become depressed. On top of the space crunch brought on by the GOT renovation, soon Bill will leave on the aquarium's annual collecting expedition to the Gulf of Maine to bring back more animals, which will further complicate the game of musical tanks.

Seeing Kali in her barrel only feeds my longing to meet an octopus in the open ocean. I have no idea how or when this will happen. My next assignment will be taking me, in just two weeks, to Niger, to document a survey of desert antelope. Nothing could be further from my watery heart right now than an ocean of sand.

But when I return from my day at the aquarium, shocking news awaits me. Al-Qaeda operatives in nearby Mali have spread to Niger, and terrorists are kidnapping visiting foreigners. The expedition is canceled. Instead of going on safari to the Sahara, I will be diving for octopus in the Caribbean.

※

The Merrimack dive shop organizes this trip every autumn to Cozumel, one of the best scuba-diving locations in the world. The island, twelve miles off Mexico's Yucatán Peninsula, gives its name to Cozumel Reefs National Marine Park, protecting 29,000 mostly pristine acres of the second-largest barrier reef in the world, in some of the ocean's clearest waters. The park boasts some twenty-six species of coral and more than five hundred species of fish, and the chance to see octopus.

"Normally, seeing an octopus is rare," says the shop's owner, Barb Sylvestre. Most other divers I spoke with agreed. My grocer, for instance, after diving around the world for twenty-five years, has seen only one—who inked at his approach. "But in Cozumel," Barb says, "we usually see a load of them on the night dive!" "A load" for a rarely spotted species might be only two or three, but still, what a thrill that would be.

※

On the first Saturday of November, I meet my fellow travelers at the Manchester, New Hampshire, airport. This year, eight people—an auspicious number, I think—are going to Cozumel: besides me and Big D, and Barb and her husband, Rob, our group consists of three other divers and one non-diving spouse. We are a buoyant, excited group, but, after delays at Mexican immigration, when we finally arrive at Scuba Club Cozumel to get ready for our checkout dive—my first in the actual ocean—I've grown stupid with exhaustion. And now it is getting dark.

In the dimming light, the equipment looks unfathomably complex and alien. I strap my BCD to my air tank sideways. Big D (so tired, herself, that she put her wet suit on inside out) helps me reposition it. I screw on my hoses—backward—damaging an O-ring. (Isn't that what made the space shuttle *Challenger* blow up?) Now I have an air leak. I haul the cylinder back to the dive shop, get a new one, attach the hoses. Wearing my mask, my lime-green fins, and my new black and pink wet suit, at last I waddle to the dock, stride purposefully over it, and plunge into the Caribbean Sea.

An alarming amount of which immediately goes up my nose.

I bob to the surface, coughing. The water tastes like a nosebleed. I gasp for "real" air by removing the regulator. Big D gives the thumbs-down signal to descend. But I can't sink!

The other divers rush to help me: One fetches more weights from the dive shop. Rob stuffs them into the pockets of my BCD. Salt water is much more buoyant than fresh, and this is why I need this checkout dive now: to get the weights right *before* I step off a boat into the ocean. But I still can't sink. Rob adds two more pounds, then four.

Now it's completely dark. I can't see anything. Water keeps pouring into my nose. Frightened by my mistakes, I can't remember how to do anything. I feel completely shot.

"It's your first night dive!" Big D says encouragingly. Somebody procures a light. Rob has now outfitted me with 12 pounds of extra weight. I follow Doris into the sea and swim through an underwater archway. For a moment I feel exhilarated, flying through the water. But I am grateful to finally haul myself out via the stepladder—except I can't, because I can't get my fins off.

Big D helps me. I check my dive computer to see how long I've been diving in the actual ocean. Was it an hour? Forty-five minutes? I look on the screen and see I went to ten feet for two minutes—not even enough to log as a dive. The rest of the time, I was bobbing at the surface, gagging and gasping.

My God, I think. What am I going to do tomorrow?

❧

The next morning, I spend half an hour in front of the mirror like a primping schoolgirl. I'm fiddling with my mask, tightening the strap, and arranging it in different configurations around my ponytail in hopes my nose won't fill up with salt water. We'll depart at 8:30 a.m. on board the *Reef Star*, a 55-foot-long Viking-hull, custom-made in the U.S. fifteen years ago, which can travel at 20 knots per hour. Our first dive of the day is a drift dive, in which we drift with the current. After we leave the boat, we won't see it again till it comes back to pick us up. We'll be far from any pier.

"We are diving at a place called El Paso del Cedral," our charismatic, barrel-chested divemaster, Francisco Marrufo, explains just before we reach our destination. It's a long, backbone-like reef, with coral heads along a ridge separating a shallow sand flat from a deeper one. "A slow current runs along the line of corals, where we may see moray eels. There may be large schools of French grunts, yellow and blue fish who grind their teeth. We may see many red snapper . . . and"—Francisco looks directly at me—"we may see octopus." Earlier, he had shared that he especially enjoys finding these animals. "If you scare them, the eyes pop up—like a person," he said. There could be four different kinds, but it can be difficult to tell them apart when each of them can assume many shapes, sizes, and colors.

The captain cuts the motor. I slip on my BCD, close the Velcro cummerbund, adjust the chest straps, de-fog my mask, put on my fins.

"Okay, let's go!" Big D says. Holding my mask to my face, I stride off the boat to follow her into the water.

My mask does not flood. I'm breathing fine. Carefully, I look down, and discover a fantasy world of colors and shapes from a psychedelic poster. Except these colors and shapes are alive: fishes, crabs, corals, gorgonians, sponges, shrimp. Corals pout like the lips of giants, and point like the fingers of skeletons. Sea fans flutter more delicately than the finest lace. The sand is New-Hampshire-snow white, the water piercing turquoise, and all around us, wild animals swim by as if we aren't there. It's like being an invisible time-traveler to another planet. Except this is the planet where I've lived for more than half a century, visiting all continents save Antarctica. And yet most of the planet has remained a distant mystery to me. Until now.

Fish are everywhere; visibility, virtually unlimited. My fear has vanished.

Almost immediately Francisco points out a five-foot moray eel, hiding beneath a ledge. It's a beautiful ribbon of velvety moss green. When it opens its mouth, I can see its pointy teeth. Scott told me there was once a moray at the aquarium who opened his jaws exceptionally wide, inviting divers to carefully scratch inside his mouth, which the gentle fish enjoyed very much. I feel as if I am meeting a friend of a friend.

Francisco is part Mayan, but he is also, I think, part fish. He slips through the water with the casual ease of a local showing us his neighborhood. So I follow him closely, keeping an eye out for Big D. We swim farther; at one point I note on my dive computer that we are at 50 feet, and my ears feel fine. Now Francisco turns to beckon us. He points at a hole beside a huge brain coral.

I see an eye, then a funnel. I hold up eight fingers, and Francisco nods. Brown mottling, white suckers. Peeling one arm away from its rock, the octopus advances, eyes tall, looking at us. Its head is only the size of a fist. It suddenly flashes red, then blanches, then shows a turquoise sheen. It withdraws all but its eyes from the hole. Then the animal peeks out farther, showing the head again, then the mantle. The funnel aims at us, then swings to the side. The gills flash white with each breath.

I could stay forever here just watching it breathe. But everyone else deserves to see the octopus too, so I move aside, inventing a new signal for Francisco: my fingertips slightly overlapping, palms toward my chest, bringing my hands in and out toward and away from my hammering heart. But Francisco already knows, has seen the rapture on my face. For more than a year and a half, since meeting Athena, since coming to know Octavia and now Kali, each time I've reached into the tanks where we have brought these creatures into our world, I've longed to enter theirs. At last, in the warm embrace of the sea, breathing underwater, surrounded by the octopus's liquid world, my breath rising in silver bubbles like a song of praise, here I am.

There follows a parade of wonders: A splendid toadfish hides beneath a rock. Once thought to live only in Cozumel, it's pancake flat, with thin, wavy, horizontal blue and white stripes, Day-Glo yellow fins, and whiskery barbels. A four-foot nurse shark sleeps beneath a coral shelf, peaceful as a prayer. A trumpet fish, yellow with dark stripes, floats with its long, tubular snout down, trying to blend in with some branching coral. Big D invents a hand sign on the spot: a fist to the mouth, holding the thumb of the other hand, with fingers up and wiggling, as if loosing the notes of a wind instrument. A school of iridescent pink and yellow fish slide by inches from our masks, then wheel in unison like birds in the sky.

I have known no natural state more like a dream than this. I feel elation cresting into ecstasy and experience bizarre sensations: my own breath resonates in my skull, faraway sounds thump in my chest, objects appear closer and larger than they really are. Like in a dream, the impossible unfolds before me, and yet I accept it unquestioningly. Beneath the water, I find myself in an altered state of consciousness, where the focus, range, and clarity of perception are dramatically changed. Is this what Kali and Octavia feel like all the time?

The ocean, for me, is what LSD was to Timothy Leary. He claimed the hallucinogen is to reality what a microscope is to biology, affording a perception of reality that was not before accessible. Shamans and seekers eat mushrooms, drink potions, lick toads, inhale smoke, and snort snuff to transport their minds to realms they cannot normally experience. (Humans are not alone in this endeavor; species from elephants to monkeys purposely eat fermented fruit to get drunk; dolphins were recently discovered sharing a certain toxic puffer fish, gently passing it from one cetacean snout to another, as people would pass a joint, after which the dolphins seem to enter a trancelike state.)

The desire to change our ordinary, everyday consciousness does not seize everyone, but it's a persistent theme in human culture. Expanding the mind beyond the self allows us to relieve our loneliness, to connect to what Jung called universal consciousness, the original, inherited shapes shared with all minds; unites us with what Plato called the *animus mundi*, the all-extensive world soul shared by all of life. Through meditation, drugs, or physical ordeal, certain cultures encourage seeking altered states to commune with the spirits of animals, whose wisdom may seem hidden from us in ordinary life. In my scuba-induced altered state, I'm not in the grip of a drug: I am lucid in my immersion, voluntarily becoming part of what feels like the ocean's own dream.

Who is to say that dreams are not real? Hindu mythology tells the story of the ascetic Narada, who won the grace of Vishnu and was invited to walk with the god. When Vishnu became thirsty, he asked Narada to fetch him some water. Narada went to a house and there met a woman so beautiful he forgot what he came for. He married the woman; together they farmed the land, raised cattle, and had three children. Then came a violent monsoon. Floods threatened to carry away the village's houses, the cattle, the people. Narada took his wife by the hand, his children by the other. But the waters were too strong and they were lost. Narada was swept beneath the waves. Washed up on shore, he opened his eyes . . . to see there, still waiting for his drink of water, Vishnu—the god who is often pictured as sleeping on a fathomless ocean as his dreams bubble forth to create the universe.

Once back on board the *Reef Star*, I pull off my mask and weep with joy.

❧

Daily, I am drunk with strange splendors: three-inch yellow sea horses with prehensile tails like those of possums; angelfish of six species, with dorsal fins trailing like bridal trains; fish with yellow lips, fish with purple tails; fish bright as parrots, fish shaped like disks; fish with elaborate patterns like chain mail or with leopard spots interspersed with tiger stripes; fish with evocative names: sergeant major. Blue chromis. Harlequin bass. Fairy basslet. Slippery dick.

One night, we dive from shore. Instantly, I lose our group in the dark and join another by mistake. Disturbed and disoriented, I swim back to the dock, disappointed to have to abort the dive. But Rob and Big D return. "Let's find you an octopus!" Rob says. Rob holds my hand and shines his light on a porcupine fish, who can blow up like a balloon when disturbed; a cowfish, with horns on the head like a steer; a flat, ghostly southern ray lying on the sand . . . and then

Rob squeezes my hand and points the flashlight at something else on the bottom. At first I think he's showing me the fat orange starfish. But just next to it, from a crack in the dead coral, something brownish red is oozing out toward us. The octopus unfurls its arms, showing its white suckers, its eyes standing up tall. Irritated by the light, it turns bright red and pours itself back down its hole, vanishing like water down a drain.

⚜

Tuesday, November 7: "Today," Francisco tells us at the dive briefing, "we're gonna dive one of the parts of Columbia: Columbia Bricks." In my guidebook, I had read about this reef at the lip of the drop-off at the island's southern tip: "Huge pillars of coral loom over white sand and slope downward on the seaward side to the successive terraces below. . . . " It's known for gigantic plate coral, gorgonian sea fans, giant sponges, huge anemones.

"Right where we start the dives are lots of bricks and one anchor," Francisco continued. "After we meet on the bottom, we're gonna swim across the shelf and come to the drop-off. There are pinnacles, some with overhangs, and then we'll go out to the wall. You may see turtles. Last week, twenty-five big dolphins came up to us from behind, toward the middle of this place. We will see sharks and stingrays, and sometimes there are ten lobsters in one place.

"We're going to dive to eighty feet. If the current goes fast, stay close to the reef. And when you come up, keep drifting."

Big D strides overboard first, and I follow. But something's wrong. At ten feet my ears hurt. I ascend slightly and try to clear them by holding my nose and blowing. It doesn't work. I see the others on the bottom. I try to descend, but the pain is too bad. I sign to Doris "trouble equalizing ears" and then sign the same to Rob.

Rob shows me some tricks: tipping the head to one side, then the other. Blowing out the nose without holding it. Rising just a little more and trying again. But nothing works, and I suppose it's no wonder: I did three dives yesterday including the deepest of my life, 84 feet, and then this morning I forgot to snort my usual decongestant spray.

Rob and I surface. "How much can this hurt and I still continue?" I ask. "Better not risk it," Rob advises. And I consider: Tomorrow is the night boat dive—the best chance of the week to see octopuses. I can't afford to miss it.

The crew helps me clamber back onto the boat and I sit miserably on a bench, my head and its uncooperative ears in my hands. I swallow a Sudafed, hoping it will clear my ears in the hour and a half between now and the next dive.

I expect the time to drag, but it goes quickly. In the sea, perhaps, time itself is slowed by the water's weight and viscosity. Even with just my hands in Kali's or Octavia's tank, time proceeds at a different pace. Perhaps, I muse, this is the pace at which the Creator thinks, in this weighty, graceful, liquid manner—like blood flows, not like synapses fire. Above the surface, we move and think like wiggly children, or like teens who twitch away at their computer-phones, multitasking but never focusing. But the ocean forces you to move more slowly, more purposefully, and yet more pliantly. By entering it, you are bathed in a grace and power you don't experience in air. To dive beneath the surface feels like entering the Earth's vast, dreaming subconscious. Submitting to its depth, its currents, its pressure, is both humbling and freeing.

A half hour later, when my friends emerge, my ears are no better. I discover Mike, another diver in our group, also had trouble with his ears, but unlike me, he completed the dive. Now he has a nosebleed and must sit out the next dive, also disappointed.

Francisco begins the briefing for the dive Mike and I will miss:

It will be at Parque Chankanaab, which is known for its lobsters and toadfish. "The second part, for me, is the best," Francisco says, "because we might see a turtle. If we do, it will be a green sea turtle."

"Like Myrtle!" Big D says to me. Today, she notes, is the day she usually volunteers at the aquarium: "I wonder if Myrtle's missing me today?"

"Fifty feet, max," says Francisco. "Be careful. The sand is like powder and easily disturbed."

And then Mike and I watch as our friends stride into the turquoise water without us.

For once, I can watch the transformation. Till now, I've been so focused on my own frantic preparations. We enter the ocean by walking overboard, shuffling on our big fins, an entry method called "performing the giant stride." It sounds stately and accomplished, but doing it, even Jacques Cousteau looked like he'd just left Monty Python's Ministry of Silly Walks. To see my friends, seasoned, graceful divers, looking so pathetically awkward and helpless, so willing and vulnerable, is a shock. In a heartbeat, the diver is reborn, swallowed into another reality, transformed from a shambling monster into a being of weightless grace. Is this what happens to the spirit at death when it flies up to heaven?

✳

Wednesday dawns, the day I'd normally see Kali and Octavia. This Wednesday is the day of the night dive from the boat, the best chance to see wild octopuses of the entire trip. There is much discussion about my ears. Mike and Rob think I might be okay, but Big D and Barb strongly feel I shouldn't attempt the first morning dive, because it will be the deepest, 70 feet. After that, there is an afternoon dive. And after that, before the night dive, a dive at twilight.

So though I come on the boat with the others, I skip the first dive, sad I'll miss the green morays, turtles, and sharks Francisco tells us are here. It's a choppy day, and the current is strong. Everyone hurries overboard, eager to get below the big waves. But something's wrong with Big D's gear. The inflator hose on her BCD is not attached properly. Two boat hands help her right the problem, fast as a pit crew, but everyone else has descended already and Big D is late. Once she strides overboard, I watch anxiously to see if she has joined the others. But because of the waves, I see nothing, not even bubbles, no evidence that my beloved instructor ever existed at all, and no proof she has joined our vanished friends. The boat moves off, to drop off another group. Big D must know what she's doing, I think, but I'm worried nonetheless.

The captain shares my concern. After dropping the other group, he turns to go back. But the water is crowded with boats and divers. Where is our group? Suddenly we see a tall, skinny orange inflatable—a "safety sausage," as it's called—and with it, Big D! Is she hurt?

She's fine, but she lost the group. "I kept lookin' and lookin'," she says matter-of-factly, "and I'm thinkin', I don't know if I'll be able to find 'em!" Big D is typically cheerful as she hauls herself out of the water and onto the deck. "I had this safety sausage for four years, and never before got to use it!"

The boat hands scan for fizz on the waves, the bubbles of the others. Ultimately they spot the group—and back down my big-hearted instructor goes, completely unfazed.

Others won't be so calm today, though. Because of the chop and the strong current, many divers had the same problem Doris had. During the second dive, which I also sat out, the water was crammed with more sausages than a German butcher shop. We rescued one of the lost floaters—an older gentleman who was quite shaken by the experience. "Usually when I surface, my boat is waiting for me!" he sputtered. But he couldn't remember the name of his vessel or his divemaster. We had room for him on our vessel because we had lost the unfortunate fellow we dubbed the Pukey Guy, after he had earlier thrown up, not over the rail as you are supposed to, but on deck, inspiring others to do the same. Later we spotted him on a different boat, which he had boarded by mistake. We eventually found the lost gentleman's vessel and retrieved the Pukey Guy, who managed to leave his weight belt on the other boat.

With so many divers lost in the daylight, I worry: What will happen on the night dive?

※

We meet at the dock at three for the twilight dive, because we have to travel an hour before we reach the site. "This place is called Delilah," says Francisco. "There's no reason to go deeper than 60 feet. It's early. It's not going to get very dark. But take a little light. Check inside the craters. This time of day, we get to see turtles going south, nurse sharks looking for a place to sleep, and parrotfish, too. If the current picks up, stay close to the reef, okay?"

I pray my ears will hold out.

I descend very slowly, equalizing repeatedly as I go, under Big D's watchful eye. At the bottom, I signal her "okay"—and notice that everyone else is watching to make sure I'm okay too. Forty minutes fly by, and even though my dive computer registers that I've dived to 90 feet, I feel no pain, only delight, as a large blue mutton snapper follows us the entire time. As the water grows darker, I feel increasingly confident. I can do this. Now all I need is for the octopuses to cooperate.

※

It's getting dark and cold. During the hour we wait topside, off-gassing the nitrogen built up in our blood, Big D and I huddle beneath a shared towel, shivering and giggling. Now I'm nervous. I think: My ears; the dark; it's night, and this is the ocean.

Francisco convenes the dive briefing. "We are almost there," he says. "Paradise: this place is called Paradiso. Of all the places in Cozumel, this is the place of the night dive. I am thinking we are going to see octopus and shark. But every night you are here is different. One night there will be lots of octopus. On the full moon, the octopus is out because he is a predator and the moon is his strobe. But lobsters stay in their hole. You may see huge crabs. You may see large squid, too. There are eels, the sharp-tailed eel who looks like a snake. You find him on the side of the reef.

"We'll meet at the surface first, at the back of the boat, and go down together. Get your light on. When you use hand signals, light your hand. And when you surface, when you come up, shine the light on your head so the boat can find you.

"I have an orange-brown light and a green one. If you see that, that's me. Okay—let's go!"

We each have two lights: a flashlight and a glow-stick on our backs. I stride in right after Rob. Because of the problems with the shore night dive, he decides to hold my right hand throughout the dive. We descend together slowly. At three feet I start equalizing. I feel a squeeze. I blow and blow and descend some more. At ten feet, I signal Rob "trouble with ears." We rise, together, a foot or two. I Frenzel. I Valsalva. I tip my head to one shoulder, then the other. That's better. I shine my light on my left hand, signaling "okay." I drop a foot, two, three. My ears squeal. But unless the pain is shattering, I am going to continue.

Finally Rob and I are on the bottom with the others. We proceed along the reef in the dark. I am so glad he is holding my hand, because I am finding it very difficult to adjust my buoyancy, use my flashlight to see my depth gauge, clear my ears and sometimes my mask, and look for animals in the small disk of light from my flashlight all at once. It feels as if I am traveling in a small capsule in outer space. Around me the darkness is heavy and enveloping. My senses have constricted and intensified to focus only on this tiny circle of light. And here it reveals a huge crab—a tall purple turret of coral—a bright blue angelfish! A school of snapper mass beneath a coral. A spiny lobster waves its antennae. Ahead there are flashes of light, like heat lightning, from my friends' cameras, contrails of light trailing from their BCDs. And then: an octopus! I squeeze Rob's hand, but he has already seen it, oozing from its hole. It's brown with white stripes, then becoming lighter as its arms boil out of its lair. Three arms walk forward, and then it turns its head, its eyes looking directly into our faces, turns green, then brown, then disappears.

Yellow coral animals are extending their feeding tentacles. Purple and orange sponges heave into view. A second octopus! Its eyes pop up, then down. The area surrounding the eye seems yellow, the pupil a slit. In an instant, it flashes speckles on its skin, a starscape, and then pours itself back down its lair.

Ahead, my flashlight reveals that Francisco is playing with a puffer fish, who for some reason allows him to gently palm its belly. But Rob is twirling his light to catch my attention. Directly below us is a third octopus. I flip head down, feet up, to observe it. This octopus is larger than the other two, and the animal doesn't seem as alarmed. Its funnel faces away from me as it crawls toward me. It flashes stripes, then dots. I feel as if it's testing me, like a scientist conducting an experiment, to see what I'll do. I want to stay, but the current is sweeping me away, and so is Rob, who must not let us get separated from the others in the dark. I feel like Dr. Zhivago having just spotted his long-lost Lara in the busy city at the end of the eponymous movie—but I am in the ocean's grip and its currents propel me forward.

Marvels flash before me in the circle of my light: a sharp-tailed eel, its tail a flat paddle, pointed at the tip. Striped grunts, so named for the grinding sound they make with their teeth. A bright blue angelfish. A huge crab. But the pressure in my ears is building. I am having trouble focusing my attention. I constantly blow out my nose, trying to equalize, but instead create a bizarre underwater sound track inside my head, squeals and bubblings accompanying the Darth Vader hissing of my regulator. If it were not for Rob's hand holding mine, I would be completely disoriented.

And then, a fourth octopus, this time on a reef wall! This one is quite small and shy, and all I see are eyes and suckers peering from a hole in the corals. My ears are screaming as Rob gives the thumbs-up sign that it's time to surface. I ascend with him slowly, like a dying soul reluctant to leave its body, and we watch the silver trail of our bubbles rising above us like shooting stars.

CHAPTER SIX

Exit

Freedom, Desire, and Escape

Upon my return, the elderly Octavia is still going strong. She is very active, swirling her suckers, turning her mouth toward the glass. Then she flips back, her body hanging below her head. She makes an eyebar, then mottles, then runs three arms across her brow. Gaping her gill opening wide as a pitcher, she inserts one arm into its entrance and then pokes the tip of her arm out her funnel, waving it like a person hailing a cab. She pulls that arm out, then sticks in another arm. Now she grows paler, expanding hugely with each breath and exhaling forcefully through her funnel. Her pupil is a fat bar, giving her an intense expression. Then she rotates her funnel, more flexible than a tongue, out of my sight. She continues to modify her coloring: Her eyebar gone, now she creates a starburst pattern. Her mottling is as rich and varied as a plush Persian carpet as she fluffs the eggs toward the back of her lair with one arm. She turns and I see that the eggs extend two feet back—there are not just thousands, but tens of thousands of them. I point them out to two children and their mom, and they gasp.

Half a flight above us, opening the top of the tank, Wilson offers Octavia first one squid, then another, in the long tongs. I watch downstairs with the transfixed children. As the octopus eagerly eats the squid, the sunflower sea star stretches the tube feet at the tip of one arm toward Wilson. "He's begging for a fish," I tell the kids. "He's got no brain, but he's not stupid. Watch!" Wilson obligingly gives him a capelin, and the star, his stomach side plastered to the glass at eye level to the kids, begins to pass the food from one thin, stalklike foot to the next. While the children watch in slack-jawed wonder, the star slowly conveys the food the full nine inches from the tip of his arm to his mouth—through which he then extrudes his stomach. "He can drool acid right out of his stomach to dissolve his food!" I tell the kids. They squeal with delight as the fish melts away like a cough drop in a person's mouth.

Kali is nearly as big as Octavia now, and the problem of where to put her is pressing. Last week, Christa and Wilson tell me, when they fed her, her arms came gushing out of the barrel with such force it was all they could do to peel her suckers away in time to keep her from escaping. "She seemed desperate to get out," Christa tells me. Today, though, Kali is not at all agitated, but seems friendly and calm, and her cold, wet embrace feels like a warm welcome.

Perhaps last week's ruckus was due to her new neighbors. A sick surf perch who shares the sump's water is being treated with Praziquantel, a drug whose effects on octopuses are unknown, so Bill moved Kali's barrel to a huge open tank with a different water source, just a few yards away. The tank in which the barrel now floats is occupied by animals he collected on the Gulf of Maine expedition: anemones; orange-footed sea cucumbers who look like Technicolor pickles; stalked tunicates, sea squirts shaped like ginger roots; and lumpfish, battleship gray, appealing and chubby, with mouths shaped in an *O* of perpetual surprise. The lumpfish is equipped with an adaptation to surf: a single suction cup on the belly, allowing it to adhere to any surface like a window decoration. And lumpfish are smart. A 2009 video records the feats of one named Blondie who, working with one of the marine mammal trainers at the aquarium, learned to swim through hoops, blow bubbles on command, hold still for veterinary skin scrapes, and swim in tight circles at the surface. In an obedience-school class my border collie Sally and I once took together, the command for this last behavior was "spin"—a trick which I, even though working with a dog breed famed for its intelligence, failed to teach her.

One of the new lumpfish seems curious about Kali. Yesterday, Bill tells me, while he was feeding the octopus, the fish came over to investigate the tips of Kali's arms.

"Maybe it makes things more interesting for her," Christa suggests. "I hope so," says Wilson. "She could use some fun."

Even without touching her neighbors, Kali can taste them. Her chemoreceptors can pick up chemical information from a distance of at least 30 yards. One researcher found that octopus suckers were 100 times more sensitive tasting chemicals dissolved in seawater than a human tongue tasting flavors dissolved in distilled water. Perhaps Kali knows her tank mates' species, their sex, their health.

Although octopuses are generally asocial with others of their kind, very little is known about their relationships with animals of other species, other than hunting prey or hiding from predators. Experts on home ceph-keeping advise the amateur against housing octopuses with other animals, because the octopus may kill and eat them. But not all interactions with tank mates are necessarily hostile. At British Columbia Waters at the Vancouver Aquarium, curator Danny Kent found that "some individuals can live with schools of rockfish for years and not eat them, while others rapidly pick off all of their tank mates in no time." One octopus who lived in the aquarium's 65,000-gallon Strait of Georgia exhibit liked to crawl up the side of the rockwork near the water's surface and trail one arm into the water column. Kent discovered the octopus was using his

arm as a fishing rod, waiting for a herring to bump into it, where-
upon he'd seize the fish and eat it.

Relationships with tank mates may be complicated. In 2000, the
Seattle Aquarium made the risky decision to house a giant Pacific
octopus in its 400,000-gallon tank with several four- to five-foot
dogfish sharks, believing the octopus would hide when threatened.
They were wrong. To their astonishment (and to the amazement
of 2.9 million viewers, when a video re-creating the incident was
posted in 2007 and went viral), the octopus instead began systematically murdering the sharks. The sharks were not missing, but
were found dead, uneaten, in the tank. This was not predation, nor
an immediate reaction to direct threat. According to the original
news reports and the text accompanying the video, the shark-killing
spree comprised a series of preemptive strikes, with the octopus taking out potential predators before the sharks even had a chance to
threaten it.

In Cozumel, I had witnessed a peculiar scene which may have
been evidence of a different kind of interspecies relationship, the
likes of which I had never seen reported before. On my last dive
of the trip, we had visited a low-profile reef with relatively few big
coral heads and long ledges and overhangs. About half an hour into
the dive, at a depth of about 30 feet, we spotted a Caribbean reef
octopus on white sand beneath a rocky overhang. I approached to
within six feet of the animal and saw to my astonishment that about
a dozen live crabs, some reddish and some greenish, with carapaces
between two to three inches long, were gathered just inches in front
of the octopus. The crabs seemed remarkably calm, considering the
predicament they were in. Some were crawling slowly, but when one
seemed to be venturing too far away from the octopus, the octopus
would extend an arm and (rather gently, I thought) sweep the crab
back in closer.

Everything about this was strange. The octopus was not bright red with excitement, as you would think it would be surrounded with a living buffet of its favorite food; it was white, with an iridescent turquoise sheen. The octopus did not seem to be using its suckers to retrieve errant crabs but instead was sweeping them toward itself with its arms. The crabs, oddly, did not scuttle. And I did not see shell or crab remains, which you usually see outside an octopus den. But perhaps this was not a den. Regardless, there were so many crabs, possibly they were standing on the exoskeletons of former companions but I just couldn't see them. The octopus looked at me briefly but then turned its attention back to wrangling crabs. It did not retreat at our approach, even as I came as close as three feet.

I had wanted to stay there longer, but the current was strong, and this was a drift dive. I ask my aquarium friends: What were all those crabs doing there? Why didn't they all run away? What was the octopus planning to do with the crabs? Was the octopus running a crab ranch? I am only half joking. I throw out another idea: Was it possible the octopus had drugged them with ink?

American marine zoologists G. E. and Nettie MacGinitie occasionally put a moray eel into a tank containing a mudflat octopus. The moray began to search for the octopus, and when it came too close, the octopus inked. The moray would continue to hunt, but it would not attack the octopus. Even if the moray actually touched the octopus, it showed no interest in attacking or eating it. The same thing happened every time.

Octopus ink, in addition to the pigment melanin, contains several other biologically important substances. One is tyrosinase, an enzyme that irritates the eyes and clogs the gills. But it may also have other effects. A 1962 article in the *British Journal of Pharmacology* reports that in experiments on mammals, the enzyme blocks the action of the hormones oxytocin (the "cuddle hormone") and vasopressin (an antidiuretic hormone that affects circulation). Fish, birds, reptiles, and invertebrates, including octopuses, have their own version of both these hormones. And, as in mammals, oxytocin has been found, in experiments with fish, to affect social interactions. If the natural level of this hormone were altered, might a normally solitary creature like a crab feel unusually calm in a crowd—even a crowd that included a major predator?

Another substance in octopus ink is dopamine, a neurotransmitter known as the "reward hormone." I had recently seen an entry mentioning dopamine in one of my favorite octopus blogs, Cephalove, an inquiry into the biology and psychology of cephalopods, started in May 2010 by then Buffalo University psychology major Mike Lisieski. Citing papers by researchers Mary Lucero, W. F. Gilly, and H. Farrington on squid ink, Lisieski speculated, "Squid ink might trick predators into 'thinking' they'd caught the squid and were eating it. . . . If a predator gets a mouthful of ink, if they can sense the amino acids that normally tell them they are eating flesh, they may behave as if they have already caught and/or eaten their prey and give up pursuit." Maybe, I suggest, the crabs were hanging out sedately because they'd been drugged into feeling happy and sated.

"I think you are reading too much into this," warns Wilson.

"What?! You think octopuses running crab ranches, corralling the crab herd by drugging them with ink is crazy?" I reply. "Then listen to this."

I relate my conversations with Peter Godfrey-Smith, a philosopher who spends his summers diving around Sydney Harbor among giant cuttlefish and octopus. He describes these encounters as "like meeting an intelligent alien."

Like humans, the cephs he met were intelligent and aware. "But look at all those neurons in the arms!" he said. "They may have a radically different style of psychological organization from us. Perhaps in octopus we see intelligence without a centralized self. If you have the design of an octopus," Peter asked, "is there a sense of self at all, a center of experience? If not, that involves imagining something so different from us it might be impossible to think of."

If there is no central consciousness, does an octopus have a "collaborative, cooperative, but distributed mind," as Peter suggests? Does it have a sense of multiple selves? Does each arm literally have a mind of its own?

It's even possible that octopuses have some shy arms and some bold arms. University of Vienna researcher Ruth Byrne reported that her captive octopuses always choose a favorite arm to explore new objects or mazes—even though all their limbs are equally dexterous. She looked at eight octopuses, all of whom would jump on prey with all their arms, curling both the interbrachial web and arms around whatever food item they would find. But they all used combinations of one, two, or three favorite arms when manipulating objects. Her team counted the octopuses using only forty-nine different combinations of one, two, or three arms for manipulating objects, when, according to her calculations, 448 combinations were actually possible if all eight arms were involved.

This could simply be an instance of handedness. Tank-bound octopuses, at least, are known to have a dominant eye, and Byrne thinks this dominance might be transferred to the front limb nearest the favored eye.

But the bold versus shy arms could be something quite different. While arms can be employed for specialized tasks—for example, as your left hand holds the nail while your right hand wields the hammer—each arm may have its own personality, almost like a separate creature. Researchers have repeatedly observed that when an octopus is in an unfamiliar tank with food in the middle, some of its arms may walk toward the food—while some of its other arms seem to cower in a corner, seeking safety.

Each octopus arm enjoys a great deal of autonomy. In experiments, a researcher cut the nerves connecting an octopus's arm to the brain, and then stimulated the skin on the arm. The arm behaved perfectly normally—even reaching out and grabbing food. The experiment demonstrated, as one colleague told National Geographic News, "there is a lot of processing of information in the arms that never makes it to the brain." As science writer Katherine Harmon Courage put it, the octopus may be able to "outsource much of the intelligence analysis [from the outside world] to individual body parts." Further, it seems "that the arms can get in touch with one another without having to go through the central brain."

"Octopus arms really are like separate creatures," Scott agrees. Not only can they grow new arms when needed; there is evidence that, on occasion, an octopus chooses to detach its own arm, even in the absence of a predator. (Tarantulas do this too—if a leg is injured, they will break if off and eat it.)

"Does one arm pull off another arm because it doesn't like its attitude?" Wilson asks with a smile.

Is this like what happens when Siamese twins fight?

Wilson says, "It's amazing how little we know about how animals live. The more you know, the weirder things get. It's really only in the last twenty years we could even be having this conversation. We're only starting to understand animals."

⚜

"I got the job!"

Christa greets me the next week in her new dark-blue polo shirt with the aquarium's iconic fish logo. To make up to its visitors for the noise and disruption of the Giant Ocean Tank reconstruction, the aquarium has hired ten new educators, to explain exhibits in greater depth and give the public a more personalized experience. "It's a temporary job till the GOT is finished," she explains, "and it's not full-time. But it's like a dream come true!" In addition to her uniform, she has also been issued a size-4 wet suit, and her first duty on her new job—which she just started yesterday—was to talk with the public as she walked Myrtle around in the waters of the penguin tray to give the plump turtle her exercise.

Or at least she *thought* that was what she was going to do. As she put on her wet suit, one of the other divers—a petite woman with red hair—had turned and asked, "Are you scuba certified?"

"Well . . . no," Christa admitted nervously. She'd been looking forward to walking with Myrtle. And now she was afraid she wouldn't be allowed to do so after all.

"If you're not certified," the red-haired diver said sternly, "you can't do this"—and then she paused and let a merry smile wash over her face—"without having a whole lot of fun!"

The mischievous diver, as it turned out, was Big D, who then showed Christa how to use a bit of lettuce to lure Myrtle into following her all over the penguin tray. Myrtle can't just be left on her own to have free run of the entire tray unsupervised, Christa explains. Because she's so big she can get wedged in between the rocks. "She loves the area near a filter, between the pipe and a wall," Christa says, and on her walks, staffers need to be careful the 550-pound animal doesn't get stuck in there.

Myrtle's exercise period lasts two hours. Each of the four sea turtles gets a personal escort during their exercise time, and each has different needs. One of the two loggerheads is blind. She was rescued in 1987 off Cape Cod in the fall with such severe hypothermia everyone thought she was dead. A worker had already started hauling her body away when someone noticed she twitched, and she was rushed to the aquarium for rehab. "That's why they named her Re-tread," Christa explains. Because she was blinded by frostbite, "when she swims toward you, you get out of her way. When she's going full blast, she'll knock you over. She's not the most graceful turtle." Another turtle, Ari, a Kemp's ridley, likes divers to bend down and lift her up in the water. She asks for this by holding her head up very high. All the divers know what she wants and rush to do her bidding. "She's got us wrapped around her little finger!" Christa says. "Or make that around the one claw on her front flipper."

❊

Even with her new job—and still working four or five nights a week at the bar—Christa makes a point of visiting with Kali with us on Wednesdays. With the surf perch cured, Bill has moved Kali's barrel back to its original position in the sump.

At about eighteen months old, Kali looks as big as Octavia now—partially because Octavia has visibly shrunken. It's ironic: Aging, shrinking Octavia, in her 560-gallon tank, wants nothing more than to stay safe with her eggs in one small corner of her den. Growing, vigorous Kali, confined to her 50-gallon barrel, is eager to explore a wider world.

Wilson wishes Kali and Octavia could switch places. But there's no way to move Octavia's eggs with her, and separating her from the eggs she tends so assiduously seems unthinkable.

"That would devastate Octavia," says Christa.

"And at this point," admits Wilson, "the eggs are a good show for the visitors, too."

One day, Octavia gave a show that we had never seen before. Christa saw it first, on a Monday afternoon during her break. The sunflower sea star, who normally stays on the opposite side of the tank, began slowly advancing along the back of the tank near the top toward the octopus. He had made it two thirds of the way across the tank when Octavia shot off her eggs, directly toward him, headfirst, arms curled and moving like a boxer. "She was only off the eggs two to three seconds," Christa said. But it was enough to impress the sea star. As he beat a slow retreat, Octavia settled back on her eggs.

Later, she did this again. Hanging in the hammock of her own arms, Octavia had just accepted a silverside from Wilson's tongs and eaten it. She dropped a second fish. Then Wilson offered a fish to the sunflower sea star. The star was about halfway across the tank with his mouth facing out toward the public. He took one fish, and had started moving it from his tube feet toward his mouth, when Wilson offered another, which he also accepted. As the two fish began their slow ride along the escalator of his tube feet toward his stomach, the sea star continued his advance across the glass in Octavia's direction. She grew more and more active as he approached, waving her arms about, suckers out, her pupils huge. She first sent one long arm out, stretching more than four feet, all the way across the tank. Then she lowered herself off her eggs, exposing hundreds of pearly egg chains. Though she was still attached by a handful of large suckers of two arms to the roof of her lair, she moved all her other arms, her interbrachial web, and her body completely off the eggs. Then she shot a powerful blast from her funnel and the egg chains swayed like curtains in a breeze. She moved her arms excitedly, suckers out and tips curled. This display lasted perhaps fifteen minutes. Finally, the sea star stopped moving toward her and reversed direction. Even without a brain to process it, he seemed to get the message. Octavia settled back down on her eggs. The movements of her arms became slower and calmer. Finally she seemed to relax.

"I think she was confused at first by the sunflower sea star," said Wilson, who had come down the stairs to watch her with us. "Then she figured out he was only eating his fish. But if the sunflower sea star had gotten any closer, I don't know what she would do." In the wild, sunflower sea stars are known predators of octopus eggs.

"She definitely wins the Mother of the Year award!" Christa said.

But despite Octavia's meticulous care, her eggs are shrinking. A few dozen have fallen to the sand below. Wilson wonders if they will eventually disintegrate. If the eggs were gone, he feels, Octavia might do just fine in Kali's barrel. And in fact, the next week, he asks Bill whether the aquarium would be willing to switch the two octopuses. But nobody wants to do it. "The eggs are too good an exhibit," Wilson tells me.

※

Some days, Kali is excited and grabby. Sometimes she plays for twenty minutes without tiring. At times like that, she might accept a fish, but not eat it right away. Instead, she wants to crawl and tug, her arms coiling up ours, sucking our skin. Sometimes she'll rise up, then abruptly sink, slacken her grip—and then, once we all relax, pull one of us with a force and suddenness that makes us all laugh at her octopus joke.

After a play session, we often rest together. She hangs at the top of the barrel, her suckers gently holding us, suspending time. Sometimes, as we watch the play of colors across her skin, it feels like we are watching thoughts flit across her mind. What is she thinking? Does Kali similarly wonder about us, as she tastes the fleeting flavors of the blood flowing beneath our skin? Does she savor our affection, our calm, our delight?

But other times, especially lately, Kali seems subdued. She touches us tentatively, her color pale. Sometimes she rises to the top to greet us, but she soon sinks, the entire bottom of the barrel covered with her arms. This scares me. Even with regular interactions with people, even with the live crabs Bill gives her to eat, can this young, growing animal thrive in a space so small and barren?

For the next few weeks, Kali's predicament dominates our Wednesday lunchtime discussions. What about shipping Kali to another aquarium with more room? Animals are constantly shuttling between aquariums. A five-foot zebra shark named Indo now swims in the penguin tray, having just arrived on loan from Pittsburgh; meanwhile, Scott is preparing to drive some of the larger, older herrings from the Temperate Gallery to an aquarium with a bigger tank in Montreal. Sending Kali away, even on loan, pains me to even mention. But would it be best for her?

No, says Scott. Bill is well aware that large octopuses are famously difficult to ship. When upset, an octopus will ink, and in a plastic travel bag with no filter for the water, a giant Pacific octopus, with enough ink to obscure vision throughout a 3,000-gallon tank, would choke on its own defense. "Plus," adds Scott, "they're more prone to stress in the first place because they're so cognizant."

We can't just build a new tank, because while the entire aquarium is in disarray, it would be difficult to justify more construction for a tank that only one individual would use for only a few months—or possibly a few weeks. Then where to put it? And even if a new tank could be built, could it be octopus-proofed? "The problem is, she could get out, and then you've got one hell of a job," says Wilson. "If you have the smallest hole, they get out. No," Wilson tells us, "Bill doesn't have any good choices. There is nothing he can do."

As distressing as Kali's situation is, Wilson is well aware that people, too, must live with space constraints. Last week, in order for the hospice facility to accommodate new patients, his wife was moved to a different room.

"Isn't that disorienting for her?" I asked.

"It's not good," he said, "but we have no other choice. Everyone is doing their best."

⚜

Wednesday, December 19: My approach to Octavia and Kali today feels particularly delicious. It's nearly Christmas. I feel it is going to be a good day. The construction noise is much louder than the classical music the aquarium is piping in to mitigate it, but the management has deployed so many educators to make up for it, the public doesn't seem to mind. It seems there is one educator for almost every group of visitors. Two divers in wet suits stand in the penguin tray waiting to answer questions; a volunteer bends down to show a first grader a model of a hawksbill turtle; other volunteers are busy showing children how to gently stroke the rays in the touch tank. The aquarium feels like the best place in the world to be.

This morning I feel compelled to sit by the Goliath grouper as I come in. He's in front of the Blue Hole exhibit. His eye swivels to notice me. I am the only one in front of his tank. We sit two inches apart, and I feel as if I could pet him like a dog. He's as big as a dog, perhaps three and a half feet long, though they can grow to eight feet. "You could put your hand in a grouper's mouth," Marion says. "And you could get it back, but it would be bloody." But it's peaceful to sit by him and have him bless me with his regard. In the wild, groupers have big, beautiful eyes that stare at visitors from the coral. They are said to be as individual as dogs and quite intelligent. Snorkelers and divers have gotten to know them as individuals.

I leave the Goliath grouper and pass the ancient fishes, the sea dragons, the Salt Marsh, the Mangrove Swamp, the herrings and jellies. Up the ramp, now curtained in fabric, I head to the flooded Amazon Forest and its separate piranha tank, to the anaconda tank with its schools of electric-blue and red cardinal tetras and busy turtles, on to the electric eel, to the New England Pond, to the Trout Stream, then another turn to the Gulf of Maine exhibit, to Stellwagen Bank, to the Isles of Shoals with its smart lumpfish and its sweet, funny-looking flounder . . . to Eastport Harbor, the goosefish and her coterie of glittering Atlantic silversides . . . to the Pacific Coastal Tidepool with its forest of giant green anemones, struck every twenty-five seconds with a crashing, bubbling wave that transforms everything like a gush of liquid lightning . . . and finally to my prize, Octavia, on her eggs, beautiful and sedate. Her eggs have a brownish tinge today. But she is tending them as carefully as ever.

I have turned on my flashlight but not taken off my coat when Anna appears, out of school for Christmas break. We hug, and seconds later, Wilson comes down. "Good. You are here," he says. "Come upstairs. Bill is moving Kali!"

Scott, Christa, and Marion wait for us in the hall.

Kali will be moving to C1, a 90-gallon tank that recently held some of the Gulf of Maine invertebrates that Bill collected on his trip. The Turner Construction crew had to make sturdy lids for tanks C1 through C3, because they needed to be able to kneel atop these tanks to get to pipes and wiring. "These lids are awesome," says Bill. The crew fashioned them from half-inch-thick Plexiglas. By adding four vise grips, Bill can clamp the lid on Kali's new tank tight enough to resist even her enormous octo-strength. It seems like the perfect solution.

Bill unscrews the top of Kali's barrel. Though she looks up, she doesn't float to the top. Bill would like to get her to go into a plastic bag and carry her the few steps across the narrow aisle to C1. "A plastic bag!" I say, taken aback. "She came here in a plastic bag," says Bill.

But Kali is having none of it. It's as if she's suspicious. Perhaps she can sense something's up.

"That's okay," says Bill. "I'll just lift up the barrel." Empty, the barrel weighs about ten pounds, but with water in it (and salt water is heavier than fresh), it will weigh at least 30 more. Kali weighs another 20. But tall, strong Bill lifts up the four-foot-tall barrel as easily as I might pick up a Kleenex. Water pours out the holes in the sides into the sump, but there's enough in the bottom to keep Kali perfectly comfortable during the six seconds it takes Bill to carry her to C1 and pour her in.

Kali rights herself in an instant and turns bright red. Immediately she begins to probe her new world with busy suckers. They flatten, suction, then slide along the glass walls of the big tank. All her arms are in motion. She concentrates her efforts on the front wall nearest to us, but also touches the sides of the tank—but not the back, facing the wall. She looks like a mime doing the "inside a box" routine, but with 1,600 suckers instead of just two palms. With the possible exception of the holding facility where she was first caught in the wild, she has never felt or tasted glass before.

Christa and Marion and Anna, Wilson and Bill, Scott and I watch, enthralled, as this young, intelligent, vigorous animal finally gets the chance to do what we've all wanted for her these long months: to explore an environment more complex and interesting than the dark barrel. Her new tank is not only larger than the old one, but it has a gravel and sand bottom, new surfaces to taste and feel, and interesting views out three sides. Another creature might be frightened by the newness, but Kali seems hungry for the wider world. She literally expands before our eyes. We have never seen her arms stretched out like this before. "She's so *big*!" says Marion. Her arms unfurled, her webbing spread, Kali seems to be soaking up sensations like a swelling sponge. She moves rapidly and purposefully, touching everything, her arms dashing about like puppies exploring the first snow, or caged birds set free. "She's so happy!" Christa cries. "Yes, very happy," Wilson says softly.

I am so glad—glad for Kali; glad for Christa in her new job; glad for Wilson, who so richly deserves some joy at this difficult time in his life; glad for Anna, whose meds were recently adjusted to free her of her tremor; glad for Marion, whose headaches are getting better; glad for Scott, whose annual trip to Brazil is next month. . . .

"Are you happy, Bill?" I ask.

"Yeah!" he says. And he clearly is happy to see his octopus enjoying her new freedom. But he's nervous too, and isn't afraid to admit it. "It's a big risk to throw her in this thing," he says. "You never really know. We *think* it's octopus-proof. But they figure out ways to do things."

I ask him what his biggest concern is. "Well, I think we have this set, but she could unscrew the standpipe in the drain." The tank drains into the sump so the water can recirculate. Kali might drain her own tank. Or she could block the pipe and flood the entire floor.

But for now, it feels like there is little room for worry amid so much joy. While her back arms continue to investigate the front and sides of the new tank, with her front arms, Kali now begins to explore the tank's porcelain lip. Wilson offers her a capelin as a distraction, which she eagerly accepts. But, since they are all such champion multitaskers, no octopus is easily distracted. Kali can feed and explore at the same time, while we can hardly process all we're seeing. Her underside is plastered to the front panel of glass, and we watch the capelin slide along her suckers, conveyor-belt-like, into her mouth. Meanwhile, more arms are curling balletically out of the tank, and Anna, Christa, and I gently wrangle them. "Tentacles *in* the tank," Anna says to her softly. Kali does not seem adamant about coming out of the tank, as she had been at times in the barrel, and we easily contain her. "She is being very gentle," says Wilson. I am almost overwhelmed with the desire to kiss one of her suckers, the way I sometimes kiss the pads of my dog's feet. But I refrain. As much as we all feel her joy as our own, Kali is, I remind myself, a large, strong, wild, nearly adult octopus. We cannot know how she might react to an utterly alien gesture from the human world.

And yet . . . Kali bobs her face to the surface and looks carefully into our eyes. Our hands respond as if summoned: We reach almost as one to stroke her head, which she seems to not merely permit, but to enjoy. She brings her eyes out of the water. Though the light is very bright, her pupils are dilated, like those of a person who's newly in love.

"Okay, let's let her rest," Wilson says. He's eager to see how the tank top goes on, how snugly it fits over the standpipe, and how he will be able to take it off for feedings and interactions in the future. Bill lifts up the Plexiglas cover and, as we urge the last tips of Kali's arms away from the edges, he places the lid atop the tank and secures it with four vise grips—and then adds 20 pounds of dive weights at each corner, just for good measure. Kali immediately reaches for the novel surface and adheres perhaps fifty suckers to the new roof of her world. In the air, her suckers stretch out an inch from holding her weight below, giving her the awkward look of a person who is hanging from the ceiling by his lips. I wonder how long she can stay there without drying out. Scott assures me, "That's what slime is for." She'll let go before she hurts herself. "She's smart, remember?"

Bill examines the lid. "The new lids for the other tanks work perfectly," he says. "But for holding an octopus . . . it might be all right. . . ." The farthest vise grip will be difficult for Wilson to reach. "It's a work in progress," says Bill. Perhaps he can add a hinge to the back? Or, suggests Wilson, the top could be cut into two sections, front and back, with the back permanently secured and only the accessible front easily opened.

Bill will consider these suggestions. "I want to get it right for the future," says Bill, "for the next octopus. I want a more permanent solution in case this problem arises again." I can feel the weight Bill has borne for these many months, the great burden he has carried, of feeling forced, by circumstances he could not predict and beyond his control, to keep a young, intelligent animal he loved in dark confinement. He says, "I don't want another octopus stuck in a barrel since May."

We watch Kali for more minutes, rapt in her octopus happiness. "This is one of those rare moments I get a warm, fuzzy feeling," says Anna. People with Asperger's often seem emotionally detached, and Anna is not given to sappy outbursts. "You get a warm, fuzzy feeling from someone who is cold and slimy," I observe. And I think: That's proof of Anna's truly great heart. And of Kali's charisma and soul.

❧

At lunch, we catch up. How has Christa's job been going? Has the new zebra shark bitten anyone? No, but a burrfish bit Christa's finger—"He follows you around and bites when he gets the opportunity. It feels like your finger is stuck in a big clamp." Marion recalls a gar who only accepted silversides of a certain shape—small and straight—otherwise, he would carry the fish over to the rays and release the food to them. Wilson tells us about an 18-inch-long shark who was added to a tank with a big grouper in it. Almost immediately, the grouper swallowed the shark, and then spat it out whole and unharmed. "But afterward," Wilson says, "that shark almost never came out. They had to feed it on a stick behind a safety net."

We are coming up against the last day of the Mayan calendar, and we joke about whether the earth's polarity will shift—and whether sharks, who can sense the magnetic field of the earth, will be affected. "Will all the great white sharks show up in Martha's Vineyard looking for dinner?" Scott suggests.

With the mention of sharks, talk returns to biting, and we again try to catalog all the creatures who have bitten Anna. She counts them off. Octopus. Piranha. Geese . . . a camel bit off some of her hair once. Scott suggests we go through the alphabet to see if twenty-six animals have bitten her. We start at the end of the alphabet. What about a zebra? No? "But a zebu sucked my finger at a small zoo," Anna offers. "Does that count?" We decide it does. "What about a yak?" Yes, at a farm, she was bitten by a yak; she was feeding it and it nipped her by mistake. What animal starts with *X*? "Xenopus," says Scott, naming a species of African clawed frog. "Yes, that bit me," Anna confirms. We go back to *A*. "*A*—what about an anteater? No, that doesn't have teeth. But one could lick her. . . . What if it licked you?"

One animal that has bitten all of us is the arowana, of which there are two in the Amazon tank. Bony-tongued and carnivorous, these long, silver fish are primitive, athletic hunters that will jump out of the water to grab prey. But today there is a new arowana, a golden Asian species, who just arrived from the Toledo Zoo, and we leave our restaurant in order to visit with it. We are eager to garner its blessing for Kali's new home, for throughout Asia, this species is revered as an icon of good fortune. Home aquarists are willing to pay $10,000 to own one. Known in China as the golden dragonfish for its large, shiny, dragon-like scales, it's the most powerful fish used in feng shui, believed to bring wealth and success, and to protect its owners from danger, accidents, ill health, and bad luck.

"It is able to interpret the language, stay focused on tasks, and display a high level of intelligence," claims the website Fengshui Mall.com. "One of the most prominent abilities that the Arowana has is being able to see bad events in the future, detecting the aura of negative energy to come," the site continues, advising that its powers are maximized if its tank is placed in a main hall.

And though we are not a superstitious lot, we might be forgiven for believing this, because of what had happened with Thor, the electric eel. Anna remembers the exact date: December 7, 2011. While his regular tank was under repair, Thor was temporarily housed in one half of a large tank behind the scenes. It had been partitioned with a three-foot-tall barrier for the safety of its other temporary residents, a lungfish and a female arowana whom Scott had raised from a baby. Electric eels are not known for leaping from the water. But Thor did, landing in the other half of the tank—where he electrocuted two of the most valuable and long-lived fish in the aquarium.

That Scott had known and loved the arowana for more than a decade made her loss particularly tragic. But worse, said Anna, "when Thor killed the arowana, he killed the good luck." Immediately after the arowana's death, Scott was beset with a string of disasters. I had known of some of them, but not until Anna and Marion recited them for me did I realize how many there were.

The night of the arowana's death, while Scott was riding the ferry home, his parents were in a car accident and his mother was hospitalized. Next, a favorite uncle fell down a flight of stairs while visiting a cathedral and died. Scott himself fell down the stairs at his home, injuring himself. His son was hospitalized with a high fever. On his annual trip to Brazil, one of the participants, a longtime friend and supporter of his work, died, and Scott was forced to spend most of the trip wrestling with the unhappy task of how to get his friend's body shipped out of a foreign country and back to the States. Scott developed a skin disease. His dog died. The bad luck, in fact, had persisted into August, when his flock of chickens was decimated by a fox and he felt forced to give away the few survivors.

The new arowana's quarantine tank is auspiciously located only a few feet from Scott's desk, just inside the hallway from the volunteers' lounge. Seeing this new, beautiful fish adds to our jubilation.

We joke with Scott that now he is invincible. Surely there will be enough good luck to flow down the hall to Cold Marine and wash over Kali in her new home.

I need to leave early, because on this snowy day I took a bus instead of driving. I had planned to catch the 2:45 p.m. home. But after another look at Kali, I hesitate. I wonder, without voicing it, if I should just stay. Maybe I should try to spend the night at the aquarium and watch Kali in her new tank.

"Will anyone check on Kali tonight?" I ask Scott.

Not only does the aquarium employ night watchmen, he explains, machine system operators make rounds of every gallery, behind the scenes and in the basement, every four hours every night, checking for leaks or flooding or problems with animals. Usually if there's something wrong they can fix it, but if not, they phone the senior aquarists. That's how Scott found out this time of year five years ago that Kathleen the anaconda was giving birth, bringing him rushing in at three in the morning.

So there is no reason to worry about Kali, no reason for me not to go home tonight to my husband and border collie, no reason to change plans for tomorrow's holiday tea with Jody and another friend—no reason to do anything but get ready for Christmas, my favorite holiday, knowing that all is now right with the world. Before leaving, Anna gives me a beautiful painting she has made of a coconut octopus. She was able to paint it because her tremor is gone. The painting will occupy a place of honor on my desk, beside the framed drawing Danny made, with the help of a computer program, of me, Wilson, Christa, and him with Kali in her barrel on his and Christa's birthday.

Marion has baked Christmas cookies for us all, and I give out baklava I have made. Earlier, Octavia eagerly ate two squid. Now we can all wholeheartedly hope for her continued long life, without fearing Kali's good fortune is cheating her. I leave the aquarium singing Three Dog Night's "Joy to the World"—"Joy to the fishes in the deep blue sea"—feeling full of octopus elation, eager for the blessings of a new year.

※

At about 11:30 a.m. the next day, I noticed Scott had sent me an e-mail at 10:51: "Can you please call me on my cell phone when you get this?"

I phoned.

"I have bad news," Scott told me. "Kali is dead."

※

I tried to piece together what had happened. That night and early morning, all had been well. An astute and reliable night watchman had last checked Cold Marine at about 6 a.m. Then at about 7:30, Mike Kelleher, assistant curator of fishes, came into the gallery as usual—and, to his horror, found Kali, pale tan, on the floor at the foot of the new tank. The tank lid was just as Bill had left it, with all four clamps and 80 pounds of dive weights on top. Due to a miscommunication, Mike believed that Kali had been transferred to Octavia's tank, and that it was elderly Octavia who had escaped, not young Kali. But he didn't hesitate for a moment. Quickly he opened the lid of the tank, returned the octopus to the water, and raced off to get the vet. As Bill was walking up the stairs to work, he ran into Mike, who told him what happened. Bill dashed to the tank, tore off the lid, and started artificial respiration—which, for an octopus, entails holding the body up so the mantle opening can be flushed with salt water from a hose. Kali's siphon was still moving, though minimally, and her body and arms changed to dark brown.

The vet came running and gave her shots of dexamethasone and atropine, to try to restart her three hearts from cardiac arrest, as well as oxytetracyline, a powerful antibiotic. For a while, everyone thought they could save her. But an hour after the injections, she had again turned tan. Although her muscles still contracted and her skin darkened on contact, Kali was dead.

Only after lunch did Christa learn the news from Scott and go to pay her respects. "There was no one in the gallery," she told me as we wept together over the phone. "Her tank was covered with a black tarp. It was pretty awful. She was very flat, but kind of laid out nicely. I looked down and didn't see her eyes. But her head was toward the front of the tank on the bottom. Her arms were toward the back. She was that iconic octopus shape. The oxygen bubbler was still bubbling. It was very strange. She was a milky white. It was such a strange way to see her. You expect to see a bright red octopus, or a brown octopus. She had pinkish white toward the edge of her tentacles," Christa said, "but she was still very beautiful."

Just as when a person dies, I needed to touch base with those who had known my lost friend. "What was your favorite day with Kali?" I asked Christa. "It was the day Danny met her, and got soaked by her," she said. "And then—since we first met her, I just constantly looked forward to the next Wednesday when we'd see her again. Danny is going to be very upset. We were planning a visit to the aquarium together soon. It's not going to be as it was. . . ."

"No," I said. "I can't believe it. I can't believe this happened. We were all so happy. . . ."

We both wanted to remember, as if the memories could summon the past to replace the unthinkable present.

"I flash back to the excitement, always, before Wilson lifts the top to the barrel," Christa said. "Will she come springing up from the bottom? I always replay in my mind her unveiling herself to us in her different various ways. It was always so exciting. And the first touches—everyone can't get their hands in with her quick enough. I'm so glad I have pictures of Kali's octopus hickeys on my arm. . . ."

I phoned Anna.

"It seems impossible," I said. "Yesterday was so wonderful!"

"I guess what I've discovered," Anna said to me, "is what you do today doesn't affect yesterday." We could not change the fact of Kali's death; but not even death could eradicate the joy of the day before. After losing a friend with whom she had shared every birthday, every success, every happiness of her youth, Anna knew: "Yesterday," she assured me, "remains perfect."

❊

I spent much of that Thursday on the phone, unable to do much of anything else. I canceled my date with my friends, who understood; of course, I could not go on with the tea, because a friend had died.

"It takes a special person to understand what it means to have a friend who's an octopus," Anna told me. She imagined the conversation she might have with friends at school: "My friend died. Her name was Kali. 'What is she, from India?' No, she's from British Columbia. The Pacific Ocean, actually. She's an octopus."

I phoned Bill and left a message of condolence, but understandably he didn't pick up or return the call. I called Wilson—as much for his opinion as an engineer as for his solace and friendship. How did Kali get out?

"There's only two ways she could have escaped," he said. "Either she lifted up the lid—I've seen octopuses escape by lifting up a lid, even a heavy lid—or she escaped through a hole." But this cover was even heavier than the one on Octavia's tank, he said, "and it was on solid." There was—and had to be—a hole in the cover, in order to admit the pipe bringing fresh seawater into the tank. And any gap not filled by the pipe, Wilson thought, no matter how small, must have provided the exit route Kali had taken.

"This is no one's fault," Wilson stressed. "Bill did the best he could with what he had. It took us years to get Octavia's tank to be relatively foolproof. I'm unhappy but not surprised. One thing I'll do is talk to Bill and see what we can figure out. But we had no choice. We had to take the risk."

Kali was extremely lucky to have lived as long as she did. Most octopuses die as paralarvae. Only two in 100,000 hatchlings survive to sexual maturity—otherwise the sea would be overrun with octopuses. "And at least we know she had a good last day," I said. "Yes," said Wilson. "She had a day of freedom. And that she got out tells you a phenomenally inquisitive and intelligent creature wanted her freedom. We know, clearly, it must have taken a lot of effort to get out. A stupid animal wouldn't do that."

"She died like a great explorer," I said. Like the astronauts who died blasting off in *Challenger*, or the brave men who perished in attempts to find the source of the Nile, penetrate the Amazon, visit the poles, Kali had chosen to face unknown dangers in the quest to widen the horizons of her world.

"Octopuses have their own intelligence that we can't match," Wilson said. "And hopefully we'll learn from our mistakes. That's the best we can do. After all," he said, "we're only human."

CHAPTER SEVEN

Karma

Choice, Destiny, and Love

Last summer, Bill completed a punishing "Tough Mudder" 12-mile obstacle course in Vermont, a gauntlet of mud, fire, icy water, 12-foot-high walls, and electric shocks. The day after the race, a benefit for wounded veterans, he got up at three in the morning and drove back to work. Yet Bill had looked better that morning than he does now, the first Wednesday since Kali's death. Haggard, Bill comes down the stairs from Cold Marine to find me in front of Octavia's tank.

We hug long and hard. We don't speak of Kali at first. Instead, we talk about his other animals. We start with the three lumpfish a few tanks down from Octavia's. One of the normally gray fish has turned orange. "That's the male, and he's orange because he's in breeding condition," Bill tells me with satisfaction. He translates the fish language: "See, he's trying to impress the females with the nesting area he's picked out." Having selected a site for laying eggs in a corner among the boulders of the exhibit, the orange male is showing it off, carefully cleaning away algae and debris by blowing on the rocks. He blows at an urchin, sending it plodding away on the tube feet between its spines. An urchin can be a hazard because it might step on the lumpfish's eggs. But so far, no eggs are in the offing. The two females, not yet in breeding condition, seem unimpressed by the male's efforts. But Bill has high hopes. Two years ago, his lumpfish bred and produced eighty babies. "They are the cutest things!" Bill says. The babies he raised come to him when he leans over the tank, looking up into his face with their round eyes, chubby cheeks, and irresistible, astonished expressions.

Together we marvel at the individuals in tank after tank in his gallery. He has cared for each one every day for the past nine years, and still they thrill him. "Look, there's my basket star," he says, as we move to the Eastport Harbor exhibit. "Unbelievable, this thing. They're gorgeous." The five-inch creature looks more like some crystalline mineral than like an animal. From a central disk like the center of a daisy, but with five radiating pairs of ridges, its five arms each divide into two equal branches, which divide again to slender, coiling branchlets more intricate than the rays of the most complex snowflake.

A few steps to our left brings us before the Gulf of Maine's Boulder Reef, a 4,000-gallon tank with 1,400 animals in it: among them, 400 red anemones, 200 sea cucumbers, 250 snipe fish, hundreds of baltemia, which look like rubbery plants but are in fact mollusks like Octavia and Kali, and the mysterious, sharklike chimera. Sinuous, ancient, and possessed of an otherworldly grace, she's part cartilage, part bone, part angel, part ghost. He got her in 2007 as a mature female, he tells me. "She's awesome," Bill says. "I just love the way she moves."

Bill's affection for his animals is as clear as the spine-tipped fin on the chimera's back. That such a meticulous, caring man has lost the most intelligent, outgoing, and beloved of them all—lost her in her healthy, vigorous, promising youth—and worst, lost her, he believes, because of his mistake—seems brutally, cosmically wrong. The line from *Hamlet*, spoken by the murdered king, comes to mind: "Our wills and fates do so contrary run / That our devices still are overthrown." Bill's sorrow sweeps over my own like a sob.

And then Wilson appears. He is holding Kali's necropsy report. Conducted just an hour after her death, the exam revealed that her eyes, arms, ink sac, colon, crop, esophagus, and immature female reproductive organs were normal. Bones from the capelin we had fed her were still in her stomach. She was huge and still growing: Her longest arm stretched four feet, four inches; her head and mantle were a foot long. Everything about her was perfect. Except that she was dead.

How did she get out? There was a gap in the cover, in back of the pipe. Bill had not overlooked this. He had covered it with a plastic tarp, and stuffed it with a bristly-feeling screening material, which octopuses don't like. But Kali was not deterred. Weighing 21 pounds and with an arm span of nearly ten feet, she had squeezed out of a hole that measured two and a half inches by one inch.

A final mystery remained: Kali had died, obviously, because an octopus cannot live for long out of water—a giant Pacific can go about fifteen minutes without sustaining brain damage. But Kali should have found available water in all directions. She had been discovered within an arm's reach of the open overflow tray of her tank, full of water that was the perfect temperature and chemistry for her. Other octopuses who have escaped seem to have purposely done so in order to enter the tanks of neighbors and eat them. Why couldn't Kali find another tank and climb in?

Though not everyone subscribes to the theory, some in Cold Marine guess that Kali might have crawled over the disinfectant mat near her tank, one of which is placed at the entrance to most of the aquarium's galleries behind the scenes. To protect the animals from disease that could be brought in on the bottoms of shoes and boots, the mat is treated with Virkon, a light-pink solution that kills viruses, bacteria, and fungi. It is also a corrosive chemical and a known skin, eye, and mucous membrane irritant. And the skin of an octopus is one giant, fantastically sensitive mucous membrane. Steinhart Aquarium's assistant curator J. Charles Delbeek has likened cephalopod skin to the lining of the mammalian gut, with the result that "levels of chemicals, nutrients, pollutants, etc. that are seemingly not toxic to other species and inverts can be toxic to cephs." One touch of Virkon might have poisoned Kali.

The irony is almost too painful to bear: Kali escaped because those who loved her most were trying to give her the best life possible, and may have died because of their efforts to protect their animals from danger and disease.

The pall from Kali's death spreads like octopus ink in water. "You've got to be kidding," is what Danny said to Christa when she gave him the news at their parents' house. At first he was confused: Elderly Octavia was the one who was supposed to die, not Kali! But then Christa explained about the transfer, and how Kali found a tiny hole to squeeze through to escape. Danny replied, "Yeah. They're smart, and they camouflage, and they're friends. . . ." And then Danny grew very quiet. Christa asked him whether he wanted her to leave him alone. "And he wanted me out of the room, which is rare," she told me. "I said, 'We'll get to meet a whole new octopus, which is great.' He said, 'Yeah, but it's not going to be Kali.' She was more than I ever expected. She brought us a great circle of friends."

Bill ordered a new giant Pacific octopus on Christmas Eve by e-mail. He promises to let me know when the new one is on the way.

❧

Eight days later, just three days into the New Year, I get the call: The new octopus is scheduled to arrive the next morning. That's a Friday, Bill's day off, and he has left Dave Wedge and fellow aquarist Jackie Anderson in charge. They invite me along to pick up the animal at the Federal Express bay at the airport.

"These pickups don't always go smoothly," says pretty, pony-tailed Jackie, an expert on jellyfish culture, as we climb into the white aquarium van, its backseats torn out to make way for aquatic cargo. One day, Jackie was sent to Logan Airport to retrieve some jellyfish arriving from the Bahamas. The trip should have been a quick errand at the start of her busy day's work. But the airline had mistakenly filled out paperwork claiming it was a domestic shipment, so there was no evidence it had cleared customs. Jackie arrived at the airport at 8 a.m. and spent the day trying to reason with the airline. With each passing hour, the chances grew that the jellies would be dangerously stressed or even die. Finally, at 4 p.m., exasperated and spent, Jackie threatened to abandon the package. The officials relented—because, she said, "They didn't want a bunch of dead jellyfish hanging around the airport."

At least the jellyfish survived. As Jackie drives, she tells the story of what happened to the cuttlefish from Japan.

An outfit in Galveston, Texas, used to breed cuttlefish for the aquarium trade. After the facility was destroyed in a hurricane, Japan became the world's major supplier. There, the animals are wild-caught. But since the 2011 tsunami wrecked the Fukushima Daiichi nuclear reactor, which leaked into the ocean, all animals caught off Japanese shores were rendered radioactive. When shipments of radioactive cuttlefish arrived at Logan airport, perplexed customs officials there kept them for three days—by which time the sensitive animals had all died. (The aquarium now arranges to have cuttlefish shipped to New York, where customs personnel are more familiar with unusual cargo, and staffers drive down to get them.)

Jackie pulls up to the airport's first FedEx bay, and Dave goes inside to inquire after our cargo. The package is waiting for us just a few bays down: a 33 x 25 x 25–inch corrugated cardboard box that was originally made to ship a 27-inch flat-screen TV. It says THIS

SIDE UP. It says RUSH. The box does not say LIVE ANIMAL. You would never suspect it contains an octopus.

Twenty minutes later, we have wrestled the 135-pound box out of the van, onto a cart Scott has wheeled to the aquarium's loading dock, and pushed it into the elevator and up to Cold Marine. Inside the box is a custom-made white Styrofoam barrel. Dave lifts the lid. An ice pack wrapped in newspaper is inside, and beneath it, knotted and sealed with a tangle of beige rubber bands, a 30-gallon bag of heavy, clear plastic, containing a top layer of pure oxygen, about ten gallons of water, and our octopus. Dave cuts through the knot so we can peer inside at the occupant.

Please, please, please, I pray silently, let it be all right.

Sitting in the water is a big, light-orange blob punctuated with white discs.

"You awake?" Dave asks the animal. We see the delicate tip of one arm curl and then twist.

Jackie sniffs the water. "He smells stressed," she announces. The water in the bag gives off the scent of geraniums. Jellyfish also smell like geraniums when distressed, Jackie explains. But that varies with the species. Stressed anemones give off a sour, salty smell.

"It's nasty-looking in there," she says, peering inside the bag. Shed sucker caps are floating in the yellowish water like fake snow in a souvenir globe. It's normal for a rapidly growing animal to shed sucker caps, though in seawater this detritus would be carried away— as would the thin ribbons of excrement at the bottom of the bag.

"Nobody's at their best after a cross-country flight," I observe, "especially if you have to travel in a bag filled with your own excrement."

"I think I've been on that airline," says Dave.

"How're you doing?" he asks the octopus. An arm waves wanly. We can't see the animal's eyes, but we can see the funnel and the opening in the mantle leading to one gill inhaling and exhaling shallowly. At least it's breathing.

Dave siphons some of the dirty water into a drain on the floor, while Jackie uses a yellow plastic pitcher to pour in some clean water from the sump. The octopus tentatively explores the pitcher with the tip of one arm.

We'd love to get the octopus out of the bag, but we don't want it shocked by a sudden change in temperature or water chemistry. Jackie sends a water sample to the lab to measure pH, salinity, and ammonia levels; Dave takes the temperature: 45°F. The water in the sump is 50°F today. As we wait, I stare into the plastic bag at the new octopus. Part of the second right arm is missing its bottom quarter. What happened? Does the octopus remember? Maybe the memory resides in the lost arm. Or perhaps the other arms know about it, but the brain doesn't.

The animal turns a deeper orange now, and I contemplate its mystery. Here is someone who was born the size of a grain of rice and miraculously survived floating helplessly among the plankton until big enough to settle on the bottom. Before me is an individual who has spent months hunting prey while avoiding the jaws that lurked everywhere—fishes, seals, otters, whales—hungry for its flesh. In its short life, the animal in this bag has already survived unimaginable adventures, made death-defying escapes, and overcome heroic odds. Had it once hidden, in its youth, inside a discarded wine bottle? Had it lost an arm to a shark and regrown it? Had it ever played with human divers, amassed a crab ranch, slipped from fishermen's gear, explored a shipwreck? And how had its experiences shaped its character?

I stare into the water and ask: *Who are you?*

❊

By the time I visit the octopus next, we know a little more about her. Yet again, we have a female. Bill has examined the third right arm, which she kept hidden from us that first day, and found suckers all the way to the end. "She's pretty feisty and active," Bill tells me. She weighs about nine or ten pounds, more than Kali did when she first came, and she may be nine or ten months old.

The shipper, Ken Wong, had procured Bill's beloved George years ago, though a different shipper had provided Octavia and Kali.

"Catching an octopus is fairly involved," Ken told me, when I called him. "They're elusive. And you've got to find one appropriate for display. Thirty- and forty-pound octopuses, you don't want. You should leave them there to breed. Then there are some that are too small, and aren't appropriate." Another problem is that, this time of year, most of the octopuses are missing from one to four arms. Lingcod, voracious predators that grow to 80 pounds, with eighteen sharp teeth, are spawning, and will bite and bully octopuses to evict them from their dens and claim the holes as their own. This is likely how our octopus lost her arm.

On his first few dives, Ken had not found a suitable octopus. Sometimes he saw no octopus at all. "Sometimes you just get skunked," he said. But Ken was determined. It took him six dives, but finally he found the octopus that would be destined for Boston.

He spotted her at a depth of about 75 feet, hiding in a rock formation, with just her suckers sticking out. Ken had touched her gently and she had jetted from her crevice—directly into his waiting monofilament net.

"The net is so soft you wouldn't feel its abrasion on your face," Ken told me. "You have to treat these animals with kid gloves. You can't yank them to the surface. You don't want to shock them." The water temperature at that depth may be more than 15°F colder than the water at the surface, so he had transferred her from the net to a closed container in about 50 gallons of water, and hauled everything slowly to the surface. She never struggled or inked.

She had lived in a 5 x 5 x 4-foot, 400-gallon tank, equipped with rocks and pipe elbows to hide in, for the past six weeks. Within the first three weeks, she learned to come to him when he slapped the water, bearing food. She especially enjoyed salmon heads and crab. She was fed on a random schedule, rather like in the wild. One day she might eat a single prawn, and two days later, she might feast on two large crabs. "She put on weight at a good clip," he told me. When he caught her, he estimated she weighed about seven pounds. Now he thought she weighed about nine.

How, then, did he entice the octopus into the plastic bag for shipping? "You have to convince the animal to get in the bag," he said. "You can't force someone that smart, with eight arms. It's not quick and easy." He drained some of the water out of the tank to ease his task, but still, it took about an hour to convince her to enter the bag.

Ken had three other octopuses at his British Columbia facility, each of which was already spoken for. One was waiting for her future aquarist to fix her tank. Another awaited resolution of a problem with quarantine. In some cases, Ken has to hold out for better weather to ship an animal. Airports close for snow or heavy fog, and he won't send an octopus out if it looks like it might be kept waiting because of weather delays.

Ken was glad for news of our new octopus. "I'm happy to hear how she's doing," he told me. "I love them all." How does he feel about capturing animals in the wild and sending them to a life in captivity? He has no regrets. "They're ambassadors from the wild," he said. "Unless people know about and see these animals, there will be no stewardship for octopuses in the wild. So knowing they are going to accredited institutions, where they are going to be loved, where people will see the animal in its glory—that's good, and it makes me happy. She'll live a long, good life—longer than in the wild."

I share all Ken told me with Bill and Wilson as we lean over the barrel, looking at the new octopus. She is a deep, chocolate brown at first, then changes to red veined with pink and brown, and finally fades to a mottled fawn color, her raised papillae flecked with white, almost like snow. "What do you think of her?" I ask Wilson.

"I think . . . she's . . . almost sexy!" he answers. "There's something about her that attracts me to her. How do you describe the feeling?" My straightforward engineer friend sounds positively romantic. "You just see something there," he says dreamily.

It sounds to me like love at first sight. Is this how he felt when he first met his wife? "Now you are getting into . . . something else!" he says, laughing.

But Wilson is clearly smitten. "The pattern, the color . . ." One of the talents that served him well in the cubic zirconia trade is Wilson's extraordinary eye for color. He can tell diamonds apart from c.z. without a jeweler's loupe. (He and his partner invented a machine that could do this, by measuring heat conductivity. They once brought it to a party—resulting in a broken engagement.) Wilson can see even more beauty in this octopus than I am capable of appreciating.

But maybe I am blinding myself to her charms. After losing Kali, I feared I might feel reluctant to open my heart to another octopus so soon. Could I keep from unfavorably comparing the new arrival to our funny, demanding, playful, affectionate Kali?

Clearly, this is not a problem for Wilson. "She's so beautiful!" he says again. And it's true. She is a gorgeous octopus: healthy, strong, glowing with color.

Christa welcomes her as well. She had observed, the first day the animal arrived, a white "bindi" on the forehead. "It's just like Kali!" Christa had said. "I think it's a good omen!"

Ever since she came, staff and volunteers had been discussing possible names. Some of Bill's volunteers, who use a red covering for the flashlight when pointing out the octopus on display to visitors, had lobbied for the name Roxanne, after the popular song by the Police about a prostitute ("Roxanne! You don't have to put on the red light"). But Bill had chosen another. He named her Karma.

Why? "Because," he says, "when I moved Kali, and she died, I was forced to get a new octopus. It was karma."

In casual Western conversation, *karma* is used interchangeably with *destiny*, *kismet*, *luck*, and *fate*. Bill had chosen the name while still in the grip of what felt to us all like a star-crossed tragedy of Shakespearean proportions. During the Elizabethan era, most Europeans believed each person's fate was predetermined, hardwired by the positions of the planets and the stars. Some people still do. But the idea of karma has a deeper, and more promising, meaning than that of fate. Karma can help us develop wisdom and compassion. In Hinduism, karma is a path to reaching the state of Brahman, the highest god, the Universal Self, the World Soul. Our karma is something over which, unlike fate, we do have control. "Volition is karma," the Buddha is reported to have said. Karma, in Hindu and Buddhist traditions, is *conscious* action. Karma is not fate, but, in fact, its opposite: Karma is choice.

※

A week later, the male lumpfish is still courting. An orange lobster is standing on his chosen lair and the male is frantic to evict it. Neither of the females has taken an interest in the nest site yet. The two swim past him, seemingly oblivious, little gray blimps with wide eyes, like surprised human babies. Bill feels bad for the male, but wonders whether he ought to add a second suitor to the tank to induce the females to breed.

Meanwhile, over in Freshwater, Killer the painted turtle has fallen in love. Unfortunately the object of his affection is not another turtle, but a pumpkinseed sunfish. Apparently he considers all the other fish in the tank a threat to his one and only mate. While courting her, he attacks any other fish who comes near, and is biting everyone's fins. While Andrew Murphy, an assistant aquarist, is explaining this to some visitors, Killer drops to the bottom of the tank and kills two minnow-like killies before the people's astonished eyes.

And while the new corals for the Giant Ocean Tank are being sculpted in studios in Charlestown, Massachusetts, and in California, a quarrel has broken out among some of the fishes in their temporary home in the penguin tray. A hogfish and a butterfly fish turned up missing chunks of tails and fins. They are removed for recovery. But who is the perpetrator? Christa reports that staff are betting it's either Barry the barracuda or Thomas the dark gray moray eel. (Polly the gentle, bright green moray is not a suspect.) Once the culprit is discovered, the staff will try to restrict his movements to a safe area of the penguin tray.

What drives these animals to make the choices they do? Why pick this mate, and not another? Why choose this route, this fight, this den, and not that one? Is this behavior random or conditioned by experience? Robotic responses to outside cues? Instinct? Do animals—or people—have free will?

Though the question remains one of the great philosophical debates of history, if free will does exist, research suggests it exists across species.

"Even the simple animals are not the predicable automatons

that they are often portrayed to be," researcher Björn Brembs of Berlin Free University said—not even fruit flies, whose brains hold only 100,000 neurons (a cockroach, by contrast, has a million). He reasoned that if these small insects were mere reactive robots, then in a completely featureless room, they would move randomly. So he glued them to small copper hooks and placed them in uniform white surroundings.

Their flights were not random. Instead, they matched a mathematical algorithm for a pattern called the Lévy distribution. This search pattern is an effective way to find food, a method also known to be used by albatrosses, monkeys, and deer, and the flies made reasonable, not random, choices, too. Scientists have found similar patterns in human behavior, in the flow of e-mails, letters, and money (and, Brembs observed, in the paintings of Jackson Pollock).

Flies even display individual variation in the choices they make. Most fruit flies typically move toward light when startled—but not all, and not with equal urgency. Harvard University researchers were surprised by the degree of individual variation the fruit flies in their laboratory showed—even among flies who were genetically identical. And like us, apparently fruit flies make choices propelled by emotions like fear, elation, or despair. Another study found that male fruit flies, dejected after their sexual advances had been rejected by females, were 20 percent more likely to turn to drink (liquid food supplemented with alcohol in the laboratory) than males who had been sexually sated.

The choices available to a complex animal like an octopus are uncountable, even in a pickle barrel. Karma now rises to the top of the barrel when I slap the water, so calm in our presence she often turns nearly pure white when we play with her. She's active, but not nearly as exuberant as Kali. She prefers to suck on us with her larger suckers, sometimes hard enough to give us hickeys that persist for twenty-four hours. When we try to interact with the tips of her arms she lets them slip from our hands. After twenty minutes or so she typically relaxes, holding us gently. But then she grabs us again, more emphatically, as if to remind us: I am strong enough to pull you in. I am gentle because I choose to be.

One weekend, though, Karma was not gentle. Andrew opened her barrel to feed her a fish, and her arms shot out to grab him. She twisted, turned bright red, flipped upside down. To his surprise, at the confluence of her arms, Andrew saw her beak, and realized she was trying to bite him.

Characteristically, he kept his cool. Now twenty-five, he's been keeping fish since he was six, managing to breed them when he was seven. (When all the fish in his tank turned up dead, he didn't cry, but instead asked his mom to borrow her scissors so he could dissect the corpses and find out what went wrong.) He is so at ease with aquatic animals that, a year ago, when he was at the convenience store across the street and felt an epileptic seizure coming on, his first thought was to get back to the aquarium and specifically, to the area behind the piranha tank—to have the seizure where he felt safe. So when attacked by a giant Pacific octopus, Andrew, who also runs a business with a partner designing and maintaining tropical fish tanks, calmly peeled off Karma's suckers and stuffed her back in the barrel. "We got off on the wrong foot—or arm," he said.

Karma's sudden dislike of Andrew seems as capricious as the female lumpfishes' continued indifference to their ardent suitor. The tireless male still hasn't given up. His nest site is impeccably clean. No algae besmirches the smooth rocks that could so safely shelter hundreds of precious eggs. No sea star or urchin dares approach his carefully guarded site. He has kept even the lobster at bay. The male swims back and forth near the top of the exhibit, almost like a pacing tiger, frantic for at least one of the females to notice and admire his real estate. Yet, still, both females ignore him. Bill hasn't given up hope either. Maybe next week, he says. . . .

I'll be missing the next episode of the lumpfish saga. For next Thursday is Valentine's Day, and, with my husband's blessing, I have a date in Seattle. I'm flying across the country to watch two octopuses have sex.

✳

The top of the 3,000-gallon, two-part tank is strung with heart-shaped red lights, its glass walls adorned with shiny red cutout hearts. A bouquet of plastic roses, tied together with red satin ribbon, floats in the water. By 11 a.m., the crowd has started to build. One hundred and fifty sixth graders have arrived by school bus. Mothers are pushing babies in strollers larger than shopping carts. Eighty-eight second graders and nineteen adult chaperones, and children as young as five from other elementary schools are here. About three quarters of the people here are kids, but there are lots of adults: one fellow, sporting a red ponytail and a black leather jacket, tells me he and his girlfriend have come every year for the past four years to spend Valentine's Day here at the Seattle Aquarium's annual Octopus Blind Date.

"It's crazy, but it's pretty amazing," says a cameraman for the Seattle NBC affiliate, KOMO. Clips he films of the event will run on the four-, five- and six-o'clock news. The Octopus Blind Date has been a regular event at the Seattle Aquarium for nine years—the jewel in the crown of Octopus Week, the biggest draw of the aquarium year. The typical winter weekday draws three or four hundred people, perhaps up to a thousand on a busy Saturday or Sunday. But a weekend during Octopus Week might bring six thousand visitors.

"It's funny to think they come to see two animals mate," says Kathryn Kegel, thirty-one, the aquarium's lead invertebrate biologist. But for her, too, even after working here seven years, it's one of the most thrilling days of the year. "The matings I've seen are such a ball of arms, you can't tell apart the individual animals." She's never missed a Blind Date during her tenure. She reckons there's "about a fifty-fifty chance they'll be interested." They may do nothing. Or one might attack the other. If this happens, she and another diver will try to separate them—if they can. "There's too many arms to do much about it, though," she admits.

One year, the female killed the male and began to eat him. Fortunately this didn't happen in front of the public, but after the aquarium had closed and the animals had been left together in the tank. And once, one octopus managed to remove the barrier separating the two tanks, and the two mated the night before the Blind Date. Now the barrier is bolted shut and tied with cable in four different spots.

With eight arms and six hearts beating as one, you might think octopus sex offers a Kama Sutra of possibilities. But the octopus is an almost staid lover compared with other marine invertebrates. Take the nudibranch, *Chromodoris reticulata*, a sea slug found in shallow coral reefs around Japan. All have both male and female sex organs and can use them both at the same time. The penis of each fits into the opening of the other, and they penetrate each other at the same time. But that's not all. After a few minutes, they both shed their penises, which fall to the ocean floor—but twenty-four hours later, they *regrow them* so they can mate over and over again.

Although there are exceptions, most species of octopus usually mate in one of two familiar ways: the male on top of the female, as mammals usually do, or side by side. The latter is sometimes called distance mating, an octopus adaptation to mitigate the risk of cannibalism. (One large female *Octopus cyanea* in French Polynesia mated with a particular male twelve times—but after an unlucky thirteenth bout, she suffocated her lover and spent the next two days eating his corpse in her den.) Distance mating sounds like the ultimate in safe sex. The male extends his hectocotylized arm some distance to reach the female; in some species, this can be done while neither octopus leaves its adjacent den.

Because octopuses are difficult to find and observe, little is known about their love lives. Much may be happening that we don't expect. Males fight over females, and they fight dirty: a male will bite off a rival's ligula and then eat it. Monterey Bay Aquarium researcher Crissy Huffard in 2008 documented an Indonesian species with a surprisingly complex mating system: She discovered that males will guard their chosen females, even though "sneaker males" sometimes cuckold them. Then in 2013 researchers reported that the strikingly beautiful, newly rediscovered larger Pacific striped octopus lives in communities of up to forty animals. Males and females cohabit in dens, mate beak-to-beak, and produce not just one but many broods of eggs over their lifetimes.

Kathryn has high hopes for this year's giant Pacific couple, Rain and Squirt. Rain, the male, weighs an impressive 65 pounds. Kathryn describes him as "a big crawler and a really mellow, easygoing octopus." He was collected in May from the waters right outside the aquarium and has grown very fast. A volunteer saw him double in size since he arrived and tells me that "he's noticeably bigger every week." He's a handsome fellow, a good shade of red. One of his larger suckers stuck against the glass of his tank is two and three eighths inches in diameter, big enough to lift more than 25 pounds. He's had his turn with various toys—he particularly enjoyed handling the squishy waffle ball the otters like to play with—but he is less interested in toys these days. Time to put away childish things. Already in the past two weeks, he left two spermatophores in his tank. They look like clear, yard-long worms; keepers at one aquarium were convinced, upon finding them in their octopus tank, that their male was suffering from an infestation of parasites. The spermatophores were proof: Rain is sexually mature, near the culmination and, soon after that, the end of his short life.

The female, Squirt, is smaller, 45 pounds, and shy. She created a den when she first went on exhibit in this new tank, which octopuses don't usually do. She's named Squirt because she does—though mostly it lands on the acrylic walls, not on passersby. She enjoys opening jars, usually at night.

The two have been reaching out to one another, tasting one another, interacting suckers-to-suckers on either side of the porous barrier separating the two tanks.

The date takes place at noon, but I score a good spot by eleven thirty. Viewed from above, the tank would look like a misshapen figure eight lying on its side, with a tiny top and bigger bottom, connected with a clear passageway, which is currently blocked with a Plexiglas barrier with small holes drilled in it. Rock walls built into the back of each tank offer each octopus at least one good hiding place. Starfish and snails cruise the sandy bottom, and greenlings and two species of canary fish swim nervously around the water. Sometimes a fish disappears, eaten.

At first, Rain is resting in the upper corner of his tank, but he then turns red and begins moving around. He returns to his corner and changes to a mottled grayish color. "I'd be petrified if I were swimming and saw an octopus like that!" a teen in a leather jacket says, with his arm around his girl. Squirt is more active in her smaller tank. She is a lovely dark orange color and has raised many of her papillae.

At 11:35, the aquarium's PA system starts playing Barry White's deep, sexy bass: "I can't get enough of your love, babe." Kathryn, now in a red dry suit, sets up a stepladder by the tank. She and fellow staffer Katie Metz will remove the bolts and cables holding the barrier in place between the two tanks, and urge Squirt through the tunnel.

"Today we're getting ready for our Octopus Blind Date," announces the emcee, introducing herself as Roberta, over the PA. "If you want to be in the first few rows in front of the tank, you can sit on the floor. If you prefer to stand, do so behind those already seated."

"Crisscross applesauce, guys," says a teacher to her second graders as the little ones settle on the floor. Behind me, the crowd around the tank is lined up twelve rows deep.

"Our octopuses are very unpredictable," continues the emcee. "I can't guarantee where they'll be. If you're having trouble seeing, our cameras will display the scene on the large screen behind the white table. I'm going to ask you not to get up and move, but stay in your spot for a few minutes to see what happens. We're going to get started in ten minutes!"

All the children scream with excitement.

The beat of the music swells. Now Roberta Flack is singing, "Baby, I love you!" To pass the time, one of the teachers shows the children how to "rave" to the rhythm. "Move your hands!" she says, like a preacher at a revival meeting. "Move them hands!"

At 11:55, Roberta again addresses the crowd, standing by Rain on the larger side of the tank. "Happy Valentine's Day, everyone! I'm going to introduce you to the animals who are having a blind date today! This is Rain, our adult male," she says, gesturing to the grayish ball of suckers resting in the upper corner of the tank, "and in the smaller tank is Squirt, our female. They're going to be introduced for the first time. These animals are very solitary. Till the very end of their lives, they don't want to meet another octopus."

Unseen by most of the crowd, which is gathered round Rain

in his larger side of the tank, Kathryn and Katie enter the water to loosen the bolts that hold the barrier in place. "How many of you have been on a blind date?" vamps Roberta. "Sometimes it works out. Sometimes it doesn't. We will have to see!"

While the divers remove the barrier, Roberta gives the crowd some species info: size, life span, growth rate. "Our diver is going to encourage Squirt to try to come and meet Mr. Rain," Roberta says.

And now we can see Squirt flowing toward us, bright red with excitement. She crawls purposefully over the sandy bottom of the tank toward Rain. He has now turned from grayish to red but is still not moving. A bright white eyespot appears on Squirt's "forehead" as she stretches her second left arm toward him, reaching within three feet of his closest arm. And then, at 12:10, she reaches a second and a third arm toward him. At her touch, Rain pours down the side of the rock wall to meet the female on the bottom.

He races into her arms. She flips upside down, giving him her vulnerable, creamy white underside. They embrace mouth-to-mouth, thousands of glistening, exquisitely sensitive suckers tasting, pulling, sucking on each other. Both of them flush with excitement.

Finally, Rain completely envelops Squirt with his interbrachial web, like a gentleman might cloak his lady with his coat on a cool night. Only a few of her suckers remain visible on the Plexiglas.

Still perched above the tank, Kathryn and Katie look down on the lovers like two Cupids. These are tense moments for the biologists. Each Blind Date runs a risk. "There's always a level of concern," Kathryn told me. "But this is what happens in the wild, and whatever happens is okay." But she and Katie know both these animals personally. They love them and don't want to see them injured; they want the mating to be a success. In the past, some females inked and tried to get away from the males, so Squirt's willingness to approach Rain is a good sign.

Now that the mating octopuses are no longer moving, the children start to decamp for their school buses. Many of the kids seem baffled. To them, human sex is incomprehensible, octopus sex unimaginable. A number of adults still hang around, watching. Two guys stand solemnly in front of the tank, arms around each other. A woman with short hair doesn't understand that what looks like one octopus is really two. "They're doing it?" she asks, puzzled. "Where's the other one?"

The two animals don't move, but Rain is becoming paler and paler. "It's a date, after all," a male voice behind me says. "There's definitely got to be communication."

"Maybe psychic communication," says a woman's voice in reply.

"Could he be hurting her?" asks another woman, worried.

"It does happen," explains Katie. "You can't control these things. But his respiration—he's breathing deeply at a comfortable rate—and her not trying to get away are good indications this is going very well."

"This is the most laid-back and gentle mating I've seen," says Kathryn.

The two animals are very still now, and at 12:35, Rain is pure white, the color of complete contentment. "They're having the cigarette now," chuckles the man in back of me.

"He's never looked this white," says a tall man whose shoulder-length hair curls out beneath a rain hat. He has been watching the scene since I got here. "Oh, it's beautiful," he says softly. "They're beautiful." The man's name is Roger, and he confides that, for the past year, he's come to the aquarium twice a week, mainly to visit the octopuses. He bought a membership to the aquarium in better days. Since then, his mother died of breast cancer, his house was foreclosed, and now he is living at Compass Center, a shelter for the homeless. He is taking photos of the octopuses for an 8-by-12-foot painting he's creating for the center, "as a gift to give back to the people who've been kind to me." At first he thought he'd paint an orca, but the octopus seems more appropriate: The compass has eight points, and the octopus has eight arms. Of all the animals in the aquarium, the octopuses are his favorites. "Going here is almost meditative," he says. "Living in the world is hard, being emotional. But being with these guys gives me peace." Recently he's been offered a home in a friend's apartment; better days are at hand. "Having the peace of being with these guys," he tells me, "has lent me the time to have something good like that happen to me."

Now that the school buses have left, most of the visitors to the tank are adults. They all seem to recognize a sweetness to the scene in front of us. "It's not the sex," explains Roger. "It's that this is the culmination of their lives." I hear no snickering or jokes. Some couples come by holding hands and stand in front of the tank like they might visit at an alcove in a church. They are looking at a blessing they, themselves, enjoy. The murmurs from the people quietly watching the animals are tinged with awe.

"Look how white he is."

"And all those bumps on his skin! He looks fluffy as a lamb."

"He looks happy."

"Yeah—content."

"They're so peaceful."

"So dear. The dear, sweet things."

"They're beautiful. Just gorgeous."

And from right next to me, I hear Roger speaking softly. "I love you, Rain," he says, his voice almost a whisper. "I love you, Squirt."

❈

The animals have hardly moved for three hours. People float by them like plankton, trailing comments like tentacles. "All the octopus's internal organs are in that thing that looks like a nose!" explains a naturalist volunteer to a five-year-old.

"Their legs are coming out of their lips!" exclaims another child.

"They mate on Valentine's Day?" says a woman to her date. "How do they know it's Valentine's Day?"

Then at 2:15 the aquarium naturalist, Hariana Chilstrom, comes over. "Moving the spermatophore to the ligula is like ejaculation," she tells me. "The ligula becomes engorged like a penis." The spermatophore is produced by an actual penis inside the mantle. One spermatophore is moved from inside the mantle to the funnel. The flexible funnel moves to the groove in the hectocotylized arm and releases the single spermatophore into the groove. The spermatophore then moves down the groove in an arch and pump action down to the ligula, the tip of the hectocotylized arm.

The mating male's heart skips a beat as the spermatophore passes to the female, whose respiration increases. Just like us. And why not? "They have the same neurotransmitters as we humans do," says Hariana.

And every octopus is different. Hariana remembers one who had "a thing" for people in wheelchairs or using canes. The octopus would come close to take a look each time a person using such a device came into view. Another was particularly interested in watching small children. Often captive land predators like tigers show such preferences, too. Captive tigers are often riveted by the sight of someone with a disability, perhaps knowing they might make easy prey. Peter Jackson, chair of the International Union for the Conservation of Nature tiger specialist group, has noted that circus tigers used to stop in the middle of a performance to stare at his child, who has Down syndrome. Zoo tigers snap to attention when my friend Liz's daughter, Stephanie, rolls by in her wheelchair. But the octopuses must have other reasons. We are not on their menu, so perhaps the metal of the chairs or canes flashes like silvery scales. Or perhaps they are simply curious because these folks move differently from the able-bodied masses.

At 2:50, Rain and Squirt have shifted slightly. The scene is a peaceful and domestic one. A couple of his suckers are plastered to her face, as if he's giving her a kiss on the cheek.

At 3:07: "It might be getting toward the end," says Katie. "They're moving apart from each other." Much of Squirt's underside is now plastered to the tank's glass, the skin on the underside of her arms pink between the white suckers. Her head and mantle, gray now, are lying on their side in his arms. A curious greenling approaches and looks at them. "She's very nervy," says Hariana of the fish. She tells me they used to have a wolf eel in the tank, too. His name was Gibson. "He was a home wrecker," she says. He lived there three years, but he'd squabble with the octopuses over the dens. Gibson would bite off pieces of the octopuses' arms, and in turn would get beat up.

After so much inaction, we are all eager to see the two animals part, and watch what they do next. "I know this is going to happen the minute I get coffee. I know it, I know it," says Hariana. We stay glued to the tank.

3:45: Rain now has some dark mottling on his light webbing. Squirt's face and eye have popped up into view, showing she's bright red. We still can't see her mantle opening. Roberta climbs up the ladder to look down into the tank but she can't get a better view.

4:05: Squirt is moving slowly upward along the wall of the tank, sucker by sucker. She is much darker than Rain, who is now a pale red. Two minutes later she stops in her tracks.

An elderly Irish couple walks by. "They're mating, Leo!" the wife exclaims to her husband in a charming brogue. Until she overheard a volunteer explaining what was happening, she tells me, she had looked at the immobile octopuses and "thought it was a cardboard cutout!" Turning to her husband, she says emphatically, "It's a very beautiful experience! Very touching actually. Very moving." Her frail husband, openmouthed and clinging precariously to his walker, seems not to understand. And yet she is radiant with eagerness to share her discovery with him, animated with the same excitement they must have felt together in the early days of their long marriage.

At 4:37 Rain begins slowly moving the tips of two of his arms. He has turned white again. Squirt is now lying on her side, her mouth and the suckers surrounding it pressed against the glass, her arms outstretched in all directions like a starburst. The largest of her suckers are the size of a silver dollar. Rain molds his arms and body around her head and mantle. His funnel begins to heave. Some of her suckers seem to be flowing, as if she's fidgeting.

At 5:03, Squirt continues her slow crawl up the side of the tank, two arms stretching up high. A third arm looks like it is petting Rain. He has one arm draped over her.

5:10: Squirt turns bright orange as the two of them abruptly jerk apart. Suddenly they unfurl in an explosion of arms and webbing. He jets to his right. She follows. She hits the floating plastic roses then and pauses at the bottom of the tank for a moment. A three-foot-long white spermatophore trails like a rope out of her mantle opening.

"The couple has separated!" Hariana speaks into her walkie-talkie to the biologist who will take the night shift. "Copy that," he answers. Squirt coughs (in octopus, this is known as gill flushing, which exposes the gray gills), turns white, then red—and the two octopuses start chasing each other around the tank.

They look like great red banners flying in the wind. She starts moving to her left, across the rocks toward the tunnel, as he moves right, back toward his original perch. She gill flushes again, then turns and heads back toward Rain. It looks as if she is chasing him from his corner. She reaches her arms out to him, and he grabs her with two of his. He starts to pull her along with him as the two head off to the left again, wrapping one arm, two arms, three arms, now four arms around each other. Then they pull apart.

At 5:23, Squirt begins to flow, her interbrachial web spread like a parachute, toward the sandy bottom, but then she gathers her arms beneath her and climbs up the glass to wedge herself into the upper corner of the tank, where Rain originally lay curled before they met. Rain, meanwhile, retreats toward the smaller side of the tank.

"I've never seen so much action postcoitally!" says Hariana.

At 5:26, they seem to have settled, like this morning, at opposite ends of the tank—but now their positions are reversed: She's in his large tank, and he's extending two of his arms into the passageway, about to enter her smaller one.

"He woke up this morning and he had a nice huge home," says the dapper, silver-haired man standing next to me, "and this female comes over and has sex with him. Now? Now he's going to get stuck with a crummy little apartment. I bet he thinks, 'I never should have gotten involved!'"

❧

The two are still in their opposite corners when the aquarium closes at 6 p.m. Night staff have no instructions to resurrect the barrier.

In the morning, when I return, the two are back in their original places. The barrier between the two has been restored. The limp white tail of the spermatophore hanging from Squirt's mantle is gone. Nobody has found it yet at the bottom of the tank, but it's done its job. Its seven billion sperm would have squirted out, into her oviduct, while the two were joined. By now, the sperm will have already have attached themselves to the walls of her spermatheca, the gland where the sperm can remain viable for days, weeks, or months—until she allows them to fertilize her eggs, at the moment of her choosing.

✳

Back at the New England Aquarium, March marks other new beginnings. All but one of the new glass panels of the Great Ocean Tank are in place; the largest of the coral sculptures, each cast from real coral, are finished and installed. Bill has left for a collecting expedition to the Bahamas to procure some four hundred of the thousand new animals that the refurbished reef, with its many new hiding places, will accommodate. Even with saws and drills screaming and the pervasive scent of glue, we can finally see the future taking shape.

At lunch one day in the cafeteria, Christa describes to us her ten-year blueprint for her and Danny's lives. "It's not easy having a twin who's different like this," she explains. "You're supposed to come into the world together, and then something happens. . . ." When she applied to colleges, she was angry and upset that Danny couldn't come with her. Now her objective is to make sure they can be together. Her dream is for this part-time, temporary job to become full-time and permanent; to make enough money to afford a two-bedroom apartment for her and Danny near the aquarium; and for Danny to work in the aquarium, too, perhaps in the gift shop. Reasoning that an advanced degree in biology might help her qualify for a better job at the aquarium, she is working four days a week here, nights at the bar, and saving money for the $20,000 tuition to Harvard's extension school, where she will earn her master's while working full-time. "It's intense," she says, "but I can do it."

Marion has been missing for a few weeks from our Wonderful Wednesdays, plagued with headaches. But one week she surprises us with happy news: She's getting married. We've met her brown-haired, bespectacled beau, Dave Lepzelter, a postdoc in biophysics at Boston University, who loves *Star Wars*, their nine pet rats, and the anacondas. They haven't set the wedding date yet, but they have selected the officiant: Scott, Marion's hero and mentor. (Neither a minister nor a justice of the peace, Scott did not bat an eye at the request. He and his wife, Tania Taranovski, had chosen evolutionary biologist Les Kaufman to officiate at their ceremony, which was held at the zoo, overseen by zebras and giraffes.)

Anna, meanwhile, is dreading her seventeenth birthday without her best friend to celebrate with her. The middle of each month is a milestone, every fifteenth an anniversary of Shaira's death. But last month was different. When she visited Shaira's grave, she finally let herself cry. "My brain can keep attacking me by making me relive terrible memories I've already been through. But now," she resolved, "I am going to fight back."

She chooses to spend her momentous first day as a seventeen-year-old with Karma and Octavia, the eels and anacondas, the chimeras and lumpfish, with Scott and Dave, Bill and Wilson, Christa and Andrew and me. Christa baked tiny cupcakes decorated with icing octopuses; I made a Bundt cake with an octopus banner mounted on toothpicks. Wilson has a special present for her—a large, dried sea horse from his vast natural history collection, amassed over decades of his travels all over the world. He continues to give much of it away, as he prepares to move from the large house he had shared with his wife to a smaller apartment. Every few Wednesdays, he brings us shells, books, corals; he donated his tiger shark jaws from Mexico to the aquarium. One weekend, with Andrew's help, he packed up his last home aquarium fish, a Lake Victoria cichlid, and its tank to go live with Christa.

Wilson's wife has moved, too. She is no longer in hospice, but is in an assisted living community. For unknown reasons, the progression of her mysterious disease appears to have halted. Her doctors don't consider her terminal anymore.

With each new interaction, the octopuses remind us of endless possibilities. Karma's severed arm has started to regrow. Her initial feistiness with Andrew has abated, and she is growing into an exceptionally calm octopus. She is unfailingly gentle with Wilson, Bill, and me. She touches us with her two front arms while applying very little suction. She pulls her face out of the water to look at me, and lets me pet her head. Quite often, she's pure white, a snowpus, as we sometimes call her—though she does change color beautifully, especially when presented with her favorite toy. She especially enjoys a purple Kong rubber toy on loan from the seals. One day she clung to it from morning till closing time, and produced a series of purple veins on the milk-chocolate background of her mantle and arms to match it.

Even though Octavia's eggs are visibly shrinking, she still tends them with inspiring diligence. She appears to have taught the sunflower sea star a lesson. Now he sticks to his customary spot, as far away from her eggs as he can get.

I can't help but think about Squirt and Rain. The Seattle Aquarium can exercise an option that the New England Aquarium cannot: Because it was built just yards from the waters in which its display octopuses are caught, it can return its octopuses to the wild at the end of their lives. (Giant Pacific octopuses cannot be released in Atlantic waters; and flying Octavia back to British Columbia Pacific at her age and size would be too dangerous, even if it were financially feasible.) Squirt and Rain were released several weeks after the Blind Date, at the same area where they had been captured.

How I wished I could have seen them released! But I did watch an Internet video of the release of another captive giant Pacific octopus called the Dude. He had lived on display at the Shaw Ocean

Discovery Centre in Sidney, British Columbia, and had been captured in those waters seven months before, when he had weighed nine pounds, the same as Karma when she arrived. Returning home, he weighed more than 50 pounds.

Four divers accompanied him, swimming beside and around their friend for a full hour.

Bright orange and magnificently adorned with large, erect papillae, the Dude used his two rear arms to stride purposefully across a muddy bottom, carrying his front arms curled backward. He paused to explore, and occasionally obscure, the video camera with his suckers. Though it's not shown in the clip, a post from one of his keepers says that the Dude also caught and ate a crab, and examined several different sites for potential dens.

"He and I have had an incredible time together," his aquarist wrote. "He has been extremely social, gregarious, and an all-around great octopus. It is sad to see an empty tank now. He will be missed! Later, Dude!" (One viewer replied sympathetically, "Sorry you had to lose your bud, but now he can find an octopus, and make more dudes.")

The affection the aquarists felt for the octopus appeared to be mutual. For the hour that they swam together, though the massive octopus could have easily escaped them, the Dude chose instead to keep his human friends by his side. Only when their tanks ran low on air did the divers reluctantly bid the Dude—"the best giant Pacific octopus in the world," wrote one—goodbye.

Watching that video, I longed to return to the ocean to watch octopuses where their choices would be as limitless as the sea. Come summer, I would have a chance to get my wish.

CHAPTER EIGHT

Consciousness

To Think, to Feel, to Know

I have entered the azure waters of Paradise—where, to my alarm, I am sinking like a stone.

Minutes earlier, I had flipped backward off the side of the boat into rolling waves. That part was on purpose. Our boat, the 20-foot *Opunohu*, is too small for divers to enter the water using the giant stride. So, on my first dive since Mexico, I've successfully performed a back-roll entry. You sit facing backward on the edge of the boat with your tank suspended behind you. Holding your mask and regulator against your face with one hand, and your hoses in front with the other, you tuck your chin to your chest and lean backward to fall into the water, head over heels—a maneuver my scuba manual described as "somewhat disorienting."

But all went well, and after I signaled to my fellow divers at the surface that I was okay, holding on to the *Opunohu*'s anchor line and passing hand under hand, we descended for about 20 feet. Everything was fine . . . until I dumped the air from my buoyancy compensation device and let go. Now I am plunging toward the bottom, upside down, like a turtle on its back, a position from which I am able to watch the white bottom of our boat recede above me like a bad dream.

Luckily, my dive buddy, Keith Ellenbogen, former scuba

instructor and an acclaimed underwater photographer, grabs my hand and halts my descent. He understands immediately what's wrong. Most countries use small, light, aluminum cylinders for scuba, but here in Mooréa, still technically part of France, divers remain loyal to the original material for the first Aqua-Lung tank, developed in 1943 by their countrymen Jacques Cousteau and Émile Gagnan—durable, but much heavier, steel. Yet in my brand-new BCD, I am carrying an additional 14 pounds of weight—less than the 17 I carried in Caribbean waters, but still too much for a small person with a steel tank.

Keith's grip gives me a chance to right myself. I'm grateful but mortified. We had buoyed each other through the twenty hours of travel from his home in New York, to Los Angeles, to Tahiti, and then the ferry ride to Mooréa, imagining this moment—when after months of anticipation, we would, at last, dive the tropical reefs of Polynesia in search of octopuses. Now that we are finally doing so, Keith—a Fulbright scholar who normally dives with the likes of Philippe Cousteau—has to drag me through the water as if pulling a sled up a hill.

I'm keen to return to the spot where, just the day before, Keith had experienced what he called "one of the most exciting moments of my life."

While Keith had been diving that day, I'd been with the rest of the scientific team snorkeling, scouting study sites in the shallows. Jennifer Mather, the leader of our expedition, doesn't dive, and doesn't need to: All her studies of wild octopus have been conducted in shallow water, where she has always found plenty of octopuses. But here in Mooréa, we were having trouble.

It was not for lack of expertise. Jennifer is one of the world's few preeminent researchers of octopus intelligence. Google "octopus intelligence" and her studies are the ones you'll see cited most.

David Scheel, fifty-one, whom I had met at the Octopus Symposium, has studied giant Pacific octopus for nineteen years in the cold, murky waters of Alaska, where he pioneered the first workable way to track octopuses with telemetry—by piercing a hole in the octopus's gill slit, like you would an earlobe to accommodate an earring, and attaching a satellite tracking tag with a locking bolt. Brazilian researcher Tatiana Leite, thirty-seven, who had completed her doctoral studies with Jennifer as one of her advisers, has found and named a new species of octopus off Brazil's Noronha island, and is in the process of describing five more. A few days into our trip, we had been joined by Keely Langford, twenty-nine. She's not a scientist, but an educator at the Vancouver Aquarium, where she is renowned for her athletic diving and swimming skills, her encyclopedic knowledge of marine life, and her eagle-eyed observations.

But even with this team of experts, our first three days of scouring the shallows had passed without our spotting a single octopus.

Even by octopus standards, our study species is a master of disguise. Because it is active by day, *Octopus cyanea* is one of the best-camouflaged octopuses in the world: University of Hawaii researcher Heather Ylitalo-Ward reports that it has one of the highest numbers of pigment cells of any octopus species. And it's also, she says, one of the smartest. In Hawaii, they often carry halved coconut shells as they walk. When the octopus travels, a coconut acts as portable armor protecting the underside from predators who lurk in the sand. When the octopus flips it over, the shell can provide a handy Quonset-hut-like shelter in an area with no suitable crevices in which to hide.

Keith certainly did not expect to find an octopus on his first dive. But he did.

With divemaster Franck Lerouvreur, Keith had departed in the boat via the channel just behind the dive center at CRIOBE, the

French research station where we're staying. Within 20 minutes, they arrived at a spot to drop anchor, where they could easily search along the barrier reef to the east of Opunohu. Even though Keith, forty-two, had been diving since he was sixteen, and has dived all over the world, he had never before seen or photographed a wild octopus. But Franck's sharp eyes had been drawn to two empty scallop shells—evidence of an octopus meal. Inches from the shells, he and Keith found a hole filled with two purplish circles, each about an inch in diameter, set in a whitish background. Above the circles, like a crown, they saw the arc of what turned out to be an arm studded with suckers. The circles, they saw, were the bulbous eyes of an octopus watching them from its den. Keith was able to take several photos before the octopus retreated.

The next day, Keith and Franck returned to the site. To Keith's delight, they found the octopus right away. And this time, the animal was not shy. It allowed him to hang around while it traveled over a roughly 50-square-foot area of reef, changing color and pattern all the while. "It was like this dude was showing me around," Keith said. "He seemed playful, and not afraid at all."

My philosopher friend Peter Godfrey-Smith and an Australian dive buddy, Matthew Lawrence, had discovered a site three hours south of Sydney that they call Octopolis, where, at a depth of about 60 feet, they have found as many as eleven *Octopus tetricus* living within one or two yards of each other. These are fairly large octopuses, with arm spans of six feet or more, and distinctive, soulful white eyes that also give the species the nickname "the gloomy octopus." Matthew told me, "I've had a couple of experiences where we were diving at this site and an octopus grabbed my hand, and took me to its den, five meters away." Once, an octopus took him on what he called "a big circuit" around the area, a tour that lasted for ten or twelve minutes. Afterward, the octopus climbed all over

Matthew and investigated him with his suckers, as if, having shown him around the neighborhood, he now wanted to explore his human guest in turn. The octopuses he met, Matthew told me, were "not aggressive—they're curious." Because he dives Octopolis regularly, Matthew is certain the octopuses there recognize him. Perhaps, he mused, they even look forward to his visits. He often brings them toys—bottles, plastic screw-apart Easter eggs, and GoPro underwater video cameras—all of which they dismantle with interest and sometimes drag into their dens.

To Keith's amazement, after giving him a guided tour, the first octopus met up with a second octopus. Keith couldn't decide which one to photograph. How can you decide which of your subjects is more photogenic, when both change color and shape before your eyes?

Keith chose to stick with the first one, who crawled around the side of a rock. As Keith was photographing it, the second octopus traveled up and over a higher rock nearby, stood up tall on its arms, as if on tiptoe, and, with what looked like keen interest, leaned toward Keith and the other octopus he was photographing. "It actively positioned itself so it could observe me," Keith said. "It was so amazing to be observed like that. In all my years photographing animals underwater—sharks, tuna, turtles, fish—I've never encountered anything that watched me like this. It was like a person watching a model at a fashion-photo shoot, or watching a pro football player at a game. Most of the time, fish observe you and notice you. But they don't look at you like this, like they are watching and learning. It was one of the most incredible experiences of my life."

Possibly the first octopus had recognized Keith, and that was why it had allowed him so close, and to stay with him so long. On his second dive, Keith had spent about half an hour all told in the company of the first octopus. Perhaps the animal will feel even more at ease if they meet for a third time. And what of the second octopus? Were the two still together? Maybe there are even more octopuses in this area. We might, I hope, make a discovery of great interest to the team.

Keith and I swim over two deep channels that run parallel to the beach. In the crystalline water, we have an excellent view in all directions. Below us, a rubble-strewn landscape tells of destruction and renewal. The reefs here were relatively undisturbed until the 1980s. Then came a plague of reef-eating starfish in 1980 and '81; hurricanes and cyclones struck the island for the first time since 1906 in 1982, and again in 1991, breaking branching corals and smothering others with runoff from heavy rains. Baby corals have now begun to recolonize the area, making Mooréa a valuable living laboratory for researchers studying reef recovery. Meanwhile, the underwater landscape of holes and crevices seems custom-made for octopuses.

We descend toward the den site, 69 feet below the surface. Holding Keith's steadying hand, breathing easily underwater, and supported by the comforting pressure of the sea, I feel free, again, to join the floating parade of beautiful, improbable lives around me. Keith points to a school of yellowfin goatfish, their chin whiskers equipped with chemoreceptors that let them taste and smell food hidden among coral and under sand. Right now these 11-inch fish sport electric yellow stripes over satiny white; but, like those of the octopus, their colors aren't static. These fish are capable of a feat that earned their Mediterranean relatives an unenviable star turn at Roman feasts. Goatfish were presented to guests live, so that diners could watch them, in their death throes, change color. Around us, teacup-size butterfly fish, citrine slashed with ebony, glide alongside their mates, demonstrating a bond they will honor throughout their lives, which may last for seven years. Beneath us, emerald and turquoise parrot fish pluck algae from coral with their beaks—actually mosaics of tightly packed teeth. Each sleeps in its own private mucous cocoon, a slimy sleeping bag secreted from the mouth, to conceal its scent from predators. Parrot fish are sequential hermaphrodites: All are born female, and later transform themselves to males.

The very existence of such creatures reminds me: Anything can happen.

Keith easily locates the octopus den. The two scallop shells are still there, exactly as the octopus had left them. But the octopus is not home. Carefully we explore a 100-foot radius around the den, an area pocked with nooks and crannies into which an octopus could melt as easily as butter into an English muffin. Perhaps Keith's octopus is out hunting, and if it's nearby, there's a chance we'll spot it.

We swim together, searching, surrounded by fish with cartoon names, in *Godspell* colors, trailing drama-queen fins. Keith points, and then fins away briefly for a photograph. Treading water furiously so as not to flip or sink, I look up: My dive buddy is surrounded by eight peaceful, four-foot blacktip reef sharks. Backlit by the sun above, all nine swimming creatures are cradled in light like a halo.

We surface too exhilarated to be disappointed. But as another precious day passes with no octopus, it's hard for me not to remember what else I'm missing. The day I arrived in Mooréa was the day of Marion and Dave's wedding. Today the transformed Giant Ocean Tank officially opened to the public, filled with its spectacular new coral sculptures and hundreds of new fish. I miss my aquarium friends, both vertebrate and invertebrate—especially after what we'd been through together earlier that spring.

※

It was remarkable: Even inside the aquarium's dim halls, far from a source of natural light and enclosed in their tanks of filtered water, so many animals seemed to sense it was spring. Even though some fish, especially from the tropics, breed year around, March's melt into April had spiked a surge in piscine sex hormones.

A male fallfish—the largest member of the minnow family in northeastern North America—started showing off to the females. Carrying pebbles in his mouth, he built a mound of gravel on the bottom of the tank, and then garnished it with a silk plant he plucked and then planted in the middle. This behavior is similar to that of the male bowerbird of Australia, who, rather than flash gaudy feathers, builds elaborate, brightly decorated sculptures to attract a mate. Though the fallfish is common to swift streams and clear lakes, its elaborate mating rituals have seldom been witnessed.

In Cold Marine, the male lumpfish has finally scored. One of the females is bloated up like a beach ball, full of eggs. Any day now, she will lay her hundreds of orange eggs in the male's rocky nesting area, which he will fertilize and then assiduously guard.

In the neighboring tank, the goosefish produced another veil.

"If I ever get married," Anna told Bill and me, "I'm going to design my veil to look like this."

"Only a little less slimy?" I suggested.

Bill countered, "No—for Anna, she'd want it slime and all."

And in Freshwater, one morning I arrived to witness a historic birth. Brendan held the lip of a rare, two-inch Lake Victoria cichlid open with one hand and gently squeezed her belly with the other. Out shot twenty-three babies, each the size of guppy fry! Each female incubates her fertilized eggs in her mouth. This species, Scott explains, is so rare it hasn't got a Latin name yet; they are virtually extinct in the wild, and to his knowledge, no birth had been recorded before in captivity. "We have probably just tripled the world's population of these fish just now," Scott said.

All the breeding had added excitement to my visits that spring. Not that I needed more—my approach to Octavia's tank was thrilling enough. Though I could have easily taken an elevator or the back stairs, I always loved the anticipation of walking up the spiral ramp, past the penguin tray full of tropical fishes, past the flooded Amazon forest, past the Suriname toads (one of whom, now fully trained for display, was always visible to the public), past the anacondas and electric eel, past the Isles of Shoals and Eastport Harbor exhibits, past the goosefish and her veil, past the velvety green anemones washed by the artificially crashing surf . . . to finally arrive in front of Octavia's tank.

Until one morning, before my trip to the South Pacific, I came in and found her left eye had swollen to the size of an orange.

At first I told myself I must be mistaken. Maybe what looked like a horrible infection was really an illusion created by the water in dim light. I turned on my flashlight. Octavia's cornea was still bulging, so opaque I could not see her slit pupil.

"Oh, there you are," Wilson said. He had been waiting for me to feed both octopuses.

"Look at this!" I cried, too distressed to even greet him. "Look at her eye!"

"Oh, no," said Wilson. "That isn't good. Let's get Bill."

Bill peered into Octavia's exhibit. By this time she had turned slightly, and we saw to our dismay that the other eye was swollen and cloudy, too, though less so. "Her eyes weren't like that on Monday," Bill said with concern.

Then Octavia began to move. Sucker by sucker, she peeled away from the ceiling and sides of her lair, loosening her grip on her precious, shrinking eggs. Finally, only a few suckers of one arm remained in contact with the mass. Her seven other arms began to meander aimlessly along the bottom.

Her action mystified us. The sea star was in his usual position, as far from her as he could get. No one was threatening her eggs. There was no food on the bottom. She seemed to be . . . just wandering.

I wondered if she were blind—but perhaps that would not matter. Octopuses who have been experimentally blinded navigate flawlessly using their senses of touch and taste. Worse, she could be in pain. (Though cooks who throw lobsters into boiling water insist invertebrates' attempts to escape are mere reflexes, they're wrong. Prawns whose antennae are brushed with acetic acid carefully groom the injured sensors with complex, prolonged movements—which diminish when anesthetic is applied. Crabs who have been shocked rub the hurt spot for long periods after the initial injury. Robyn Crook, an evolutionary neurobiologist at the University of Texas Health Science Center, finds octopuses also do this, and are more likely to swim away or squirt ink when touched near a wound than when touched elsewhere on the body.)

"What is happening, Bill?" I asked helplessly.

He watched the old octopus for a few moments. Her movements seemed restless and disoriented. Her mantle was throbbing. Her body looked like the embodiment of a big headache.

"This," he said to us sadly, "is senescence."

In her old age, Octavia's tissues were simply breaking down. The previous weekend, I had seen this in my neighbor, who is ninety-two. She was thinner, dimmer, frailer. Her delicate skin bruised easily. She mentioned having seen an elephant on the lawn. Body and soul, she seemed to be deliquescing like fallen fruit.

When other octopuses go through this phase, "they sort of wander around," Bill said. "They get white patches. I haven't seen this with the eye before, though."

Again I felt the panic that had risen in my chest that night the previous August, when Octavia's body had looked like a bloated tumor, and Wilson and I both thought she was dying. But now, it seemed, the time we had dreaded had finally arrived.

"What should we do?" I asked.

There is no cure for old age, no treatment for an octopus with senility. "I like finishing up the natural process on exhibit," said Bill. "But that's not always possible. . . ."

Was there anything that would make Octavia more comfortable at the end of her long life? Should she be moved to the cozy safety of the barrel? Wild females on eggs often wall up their dens with rocks. The barrel re-created that situation more fully than the exhibit tank with its big front window.

Moving Octavia would also free up the exhibit for young Karma. She was outgrowing the barrel, as Kali had. She appeared to have given up on trying to get out. At our slap on the water, she would come up to the surface to greet us, a dark reddish brown, eat her food, and then sink to the bottom of the barrel and turn white. She was a sweet, gentle octopus, but we wondered whether it might be healthier for her to be more active.

Wilson strongly felt Octavia and Karma should be switched. Andrew and Christa were horrified at the suggestion. Interestingly, the young people were more concerned about the old octopus, and the elder was arguing for the welfare of the young one. "Take her off exhibit, away from her eggs?" Christa said. "That would devastate her!" Andrew was afraid it would kill her to move.

"But that's what we do with old people with dementia," I observed. "When they go senile, we take them off exhibit." Wilson laughed, a little sadly. "I never thought of it this way," he said, "but it's true." People with dementia can't safely negotiate the wider world. Many seem calmer in a smaller, simpler space. But what about an octopus?

We owed Octavia repose in her old age. Karma, too, of course, richly deserved the best life we could give her. But we knew Octavia better than we knew Karma. Octavia had enriched our lives since she arrived in spring of 2011, then already a large octopus with knowledge of the wild ocean, an octopus who understood camou-flage better than any other Bill or Wilson had ever met. Shy at first, she had opened up to us, and we'd won her over. I remembered so well the first time she had briefly extended the tip of one arm to my friend Liz's finger, and both withdrew; I remembered the first time she chose to interact with me—and nearly pulled me into the tank. Octavia had made us all laugh when she surreptitiously managed to seize a bucket of fish unseen, while no fewer than five people were watching her. Octavia's touch had eased Anna through the agony of losing her best friend. We had a history together. For sharing with us her surprising and revelatory life, we owed Octavia comfort and respect at its end.

We were all tortured by the uncertainty. Nature offered no ad-vice; her model is not kind. In the wild, Octavia surely would have been dead by now. Even if she had survived long enough to lay eggs and see them hatch, were she wild, she would have spent her last days wandering, alone, starving and senile, to be eaten by a predator or scavenged, like Olive, the octopus off Seattle's coast, by sea stars.

We humans changed that natural path when we took Octavia from the ocean. Because of human intervention, Octavia never got to meet a male to fertilize her eggs. Despite her assiduous care, she would never see her eggs hatch. But we had fed her and protected her; we had provided her with marine neighbors, interesting views, and entertaining interactions with people and puzzles. We had kept her from hunger, fear, and pain. In the wild, virtually every hour of every day would have brought the risk that a predator would bite off part of her body, as had happened to Karma, or that she would be torn limb from limb and eaten alive.

Since laying her eggs, Octavia seemed no longer to want our touch or our company, but she still, at least, seemed to enjoy our food. So Wilson offered her three squid. She seized the first in her left front arm, but dropped it to the bottom of the tank, where it was eagerly devoured by an orange sea star. Wilson placed the second squid directly in her mouth, where she held it for a moment, then let it go. She also dropped the third.

If Bill moved her, Octavia might feel rescued from the confusion of too much space and too many options. Or she might fight with her last bit of strength to defend her eggs. But in the absence of any cues that the eggs were alive after all these months, perhaps she was now forgetting about them. We didn't know. Nobody even knew whether Octavia could *be* moved.

Bill wasn't sure what he'd do, but no matter what his choice, the decision would be agonizing.

※

"Don't worry." Jennifer's voice, coming from beneath the mosquito net cloaking the bed across from mine in our shared room at CRI-OBE, greets me at first light. "We'll find octopuses. I don't know how many. I don't know how good the data will be. But we'll find octopuses, know that. We've got bloodhounds. These people are really good."

I haven't said a word, but Jennifer knows what I'm thinking. Field science is, by nature, unpredictable. I had learned this on other expeditions, too. We did not see any snow leopards in Mongolia; I saw a tiger exactly once in four trips to India's mangrove swamp, the Sundarbans. Sometimes your study animal does not materialize. Still, you can often accomplish a great deal: We had collected leopard scat for DNA analysis in Mongolia, and I had studied a lot of tracks, and amassed many local stories, in India. But here in Mooréa, we really must actually meet octopuses—because we need to administer personality tests to them to carry out our study.

Jennifer had created a personality check sheet to measure

whether an octopus was bold or shy. Writing in pencil underwater on a plastic dive slate, we were to record how the animal reacted to different situations. What does the octopus do when you approach it? Does it hide, change color, investigate, ink? What happens when you touch the octopus gently with a pencil? Does it jet from its burrow? Retreat? Grab the pencil? Aim its jet at the intruder? Just watch?

Our study aims to test three hypotheses about what these octopuses eat and why. David, a behavioral ecologist, suspects octopuses prefer big crabs, but will eat a wider menu if they can't find them. Tatiana, a marine ecologist, predicts that octopuses who live in more complex environments will eat a more varied diet. And Jennifer is testing for the effect of personality on food choices. She reasons that, much like many confident, intrepid people, bold octopuses may make more adventurous diners. To find out, we'll collect and identify the prey remains around each octopus den.

Jennifer had developed the personality test over many years—in spite of many arched eyebrows from skeptical colleagues. She's sixty-nine, and began her career at a time when few scientists believed animals had personalities—or that women made capable field scientists. That's why she trained as a psychologist. At Brandeis University, she studied human sensory-motor coordination, specifically eye movements, for her PhD; she later went on to study the eye movements peculiar to people with schizophrenia. But she was fascinated by cephalopods and set up tanks for *Octopus joubini* in the basement of Brandeis's Psychology Department, and began cataloging octopuses' movements and how they used their tank space.

"So when I started asking deeper questions about octopus than just 'what is it doing?' I turned to psychology," Jennifer tells me as sunrise parts the clouds over the jungle-clad volcano out our dormitory window, and roosters start to crow. "I'm quite aware octopuses don't have a mother complex—Freud is no help! But I am also aware that in animals, as well as people, there is an inborn temperament, a way of seeing the world, that interacts with the environment, and that shapes personality. There's nobody else doing what I'm doing. It may be weird, but it's unique."

Once overlooked or dismissed outright, Jennifer's work now is respected and cited by cognitive neuroscientists, neuropharmacologists, neurophysiologists, neuroanatomists, and computational neuroscientists—including a prominent international group of whom gathered at the University of Cambridge in England in 2012 to write a historic proclamation, the Cambridge Declaration on Consciousness. Signed by scientists including physicist Stephen Hawking in front of 60 *Minutes* cameras, it asserts that "humans are not unique in possessing the neurological substrates that generate consciousness" and that "nonhuman animals, including all birds and mammals, and many other creatures, *including octopuses* [italics added], also possess these neurological substrates."

No one knows octopuses like Jennifer does. If she says we'll find octopuses, I have to trust that we will.

We head out that morning to snorkel at one of the potential study sites we scouted earlier, an area with a gentle, sloping bottom that comes to a steep drop-off, boasting both live and dead coral, a hard top, and numerous gullies. While the others search the shallows, David and I swim off to the deeper area. David almost immediately finds octopus evidence: two crab claws piled on a flame scallop shell as carefully as a stack of plates piled in the kitchen sink after dinner. "A den, but no octopus," he observes. "But I declare this site very promising."

I feel like I've won the lottery. Jennifer seems most interested in searching in only three or four feet of water. But I find the shallows difficult. There we constantly plow our lips, foreheads, and chins into big clots of brown, bristly algae called *Turbinaria ornata*; at every turn, I fear I'll scrape my nipples off on the jagged skeletons of dead coral. I'm afraid I'll kick one of the few living corals, or squash a sea cucumber, or, God forbid, impale myself on the tall, black, poisonous spines of one of the sea urchins we see everywhere, or a deadly stonefish, which we don't see, because they are exactly the color of sand. (Their sting can kill you, but not before inflicting pain so unbearable that victims beg doctors to amputate the afflicted body part.)

Swimming here, in deeper water, is pure delight. All around us, fish dazzle and shimmer with iridescent stripes, glowing eyes, fiery orange bellies, black masks, Jackson Pollock spots. A hawksbill sea turtle swims beneath us, oaring the water with its winglike, leathery front flippers. More blacktip sharks slide by, weightless as dappled light. Beneath us, the bottom, flecked with blue and yellow living corals, offers what seems like countless crevices ideal for octopus.

David teaches me to free-dive. Holding your breath, you dive below to investigate a den site, then emerge blowing water from your snorkel like a spouting whale. He's found more than ten piles of food remains, so many shells and carapaces he's stopped collecting them in the lidded bucket attached to his weight belt. Inspecting coral crevices with his waterproof flashlight, David finds the evidence everywhere: Shells are stacked up one atop the other, with crab claws resting on top, like spoons in a bowl. "Nobody else is going to leave these in a pile!" he says. "The octopus must have just stepped out." In fact, by midmorning, he's located at least three octopus dens. But none of the occupants is in.

Overhead, we notice the sky is bruised with the dark clouds of a gathering storm. Reluctantly, we turn toward shore to rejoin the others. In the distance, we can see them waving at us. We swim faster to catch up. Jennifer pulls her snorkel from her mouth. "I'm looking at an octopus!" she announces, and plunges her face back into the water.

By the time I find the octopus, all I can see in the cavern into which it has retreated are some white suckers along one bluish arm. But there's more good news: This is the second octopus of the day. Tatiana had found one in the first ten minutes of their foray. The animal had been out hunting, its arms and interbrachial web spread across a shallow gully, its skin an iridescent blue-green. When it saw her, it turned its head brown first, and then its arms, and then poured itself down its hole.

The clouds now hiss rain down on us, loud as sizzling grease. The water is not a good place to be when lightning threatens, so we decide to head back to CRIOBE. As Tatiana sits down in the foot-deep water to remove her fins, David takes one last look. Right beside her, he notices a pile of shells beside a rock, a hole—and the suckers of a third octopus.

On subsequent days, we investigate more sites, mostly without octopus sightings. Still, by the end of our first week, we have located six octopuses at three different study sites. We've gathered and identified hundreds of prey items; we've collected thousands of data points from the habitat. I feel deeply attached to my new friends and, profoundly grateful for the success of our expedition, I want to give thanks. So that Sunday, when the team takes a day off, while the others sightsee and birdwatch, I go with Keith to the octopus church.

In the village of Papetoai, just a short drive from CRIOBE, there was once a temple dedicated to the octopus, the guardian spirit of the place. To Mooréa's seafaring people, the supernaturally strong, shape-shifting octopus was their divine protector, its many reaching arms a symbol of unity and peace. Today, a Protestant church occupies that site. Built in 1827, the oldest church in Mooréa still honors the octopus. The eight-sided building nestles in the shadow of Mount Rotui, whose shape, to the people here, resembles the profile of an octopus.

Taking seats in the back, Keith and I are the only foreigners to join the packed congregation of about 120 people. Almost everyone around us has a tattoo; many of the women wear elaborate hats made of bamboo and live flowers. The minister wears a long, waist-length garland of green leaves, yellow hibiscus, white frangipani, and red and pink bougainvillea; the women in the choir are adorned with headdresses of flowers and leaves. When the choir sings, their voices ring deep and sonorous, like a chant coming from the sea itself. The front of the church faces the ocean, and the sea breeze blows through the open windows like a blessing. "This is like going to Atlantis," Keith whispers.

The service is conducted in Tahitian, a language I don't understand. But I understand the power of worship, and the importance of contemplating mystery—whether in a church or diving a coral reef. The mystery that congregants seek here is no different, really, from the one I have sought in my interactions with Athena and Kali, Karma and Octavia. It is no different from the mystery we pursue in all our relationships, in all our deepest wonderings. We seek to fathom the soul.

But what is the soul? Some say it is the self, the "I" that inhabits the body; without the soul, the body is like a lightbulb with no electricity. But it is more than the engine of life, say others; it is what gives life meaning and purpose. Soul is the fingerprint of God.

Others say that soul is our innermost being, the thing that gives us our senses, our intelligence, our emotions, our desires, our will, our personality, and identity. One calls soul "the indwelling consciousness that watches the mind come and go, that watches the world pass." Perhaps none of these definitions is true. Perhaps all of them are. But I am certain of one thing as I sit in my pew: If I have a soul—and I think I do—an octopus has a soul, too.

There are no crucifixes or crosses in this church—only carvings of fish and boats, which make me feel free and forgiven. Riding the rolling waves of Tahitian vowels, I'm transported by the pastor's sermon: to the Gilbert Islands, where the octopus god, Na Kika, was said to be the son of the first beings, and with his eight strong arms, shoved the islands up from the bottom of the Pacific Ocean; to the northwest coast of British Columbia and Alaska, where the native people say the octopus controls the weather, and wields power over sickness and health; to Hawaii, where ancient myths tell us our current universe is really the remnant of a more ancient one—the only survivor of which is the octopus, who managed to slip between the narrow crack between worlds. For seafaring and coastal people everywhere, the octopus's transformative powers and elastic reach connected land and sea, heaven and earth, past and present, people and animals. Facing the ocean in an eight-sided church, drenched in blessings, immersed in mystery, my natural response, even on an expedition in the name of science, is to pray.

I pray for the success of our expedition. I pray I'll finally get to see more than just some suckers under a rock. I pray for my husband, my dog, my friends back in the States; I pray for the Giant Ocean Tank—please, God, don't let it leak!—and for my friends at the aquarium. And I pray for the souls of the octopuses I have known; those who are alive, and those who have died, but whom I will never forget.

※

After I had left the aquarium, Octavia's left eye got worse, and her right eye was cloudy. Especially if she were senescent, and not in her right mind, in a tank full of other animals and rough surfaces, the chances she might further injure herself were great. And by Thursday morning, another factor had arisen that Bill was forced to consider.

At about 10 a.m., his glance was drawn to movement in Karma's barrel. Without opening the top, he looked down on a sight he had never before seen: an octopus hanging upside down at the surface, her black beak clearly visible, persistently chewing on the mesh-like plastic conweb across the top of the lid.

Karma had already severed some of the brand-new cable ties that held the mesh onto the screw top. When Bill saw this, he understood why he'd had to replace some of the ties after Kali died. Now he realized that the damage to the ties had not just been the result of normal wear and tear: Kali, like Karma, had been systematically gnawing through them in an effort to escape.

"I was nervous," Bill told me. "I still didn't want to move Octavia." He was afraid he might injure her; he was afraid he wouldn't be able to catch her at all. He had never moved a live octopus out of the exhibit before. "But Karma gave me no choice. Octavia gave me no choice."

Bill spent the rest of that Thursday moving around fish: he transferred some redfish from Eastport to the Boulder Reef tank to make room for some rainbow smelt to move from behind the scenes to the Eastport tank, so that some new sea stars from Japan could occupy their vacated tank. He moved two small redfish, two snailfish, and one radiated shanny from behind the scenes to the Eastport tank, freeing their tank for a new, small red octopus, the size of my hand, who had just arrived. Bill chose to try to move Octavia and Karma after the public had left, because he wasn't sure how it would go.

Luckily, Bill's Thursday volunteer, Darshan Patel, twenty-nine, was there to help. Together they lifted Karma's 50-gallon barrel out of the sump and placed it on the floor. Bill propped open the lid to Octavia's exhibit. While Darshan watched from the public side, Bill used a soft, deep-bodied mesh net with a metal handle to try to scoop Octavia up from her corner. At the touch of the net, Octavia tucked up deeper into her corner; Bill couldn't reach her from that angle. So the two men switched positions, and Darshan went upstairs to make sure Octavia didn't try to escape the open tank while Bill went downstairs to assess the situation.

To give them more room to maneuver, they needed to remove an additional portion of the cover, which was bolted to the tank. Darshan donned waders so he could stand in the tank while Bill worked from above. Octavia's cloudy eyes swiveled to follow him as he worked.

Darshan is five foot ten; the water came up to his waist, but when he bent over, cold water poured inside his waders. When the lid came off, Octavia began to move toward the back glass that separates her tank from the wolf eels. Bill worked a net above; Darshan used a net, plus his free hand, while in the water. "We were being gentle, trying to guide her into the net," Bill told me. But Octavia evaded them again and again. Even with four of her arms in the net, she was holding on to the rocks with four others. As Darshan pursued her, she poured two of her arms and half her body into a crevice and refused to let go. "For an old octopus, she was still super-strong," Darshan said. "You have to have respect for her suckers. It's crazy how strong they are."

It was obvious this wasn't going to work. So while Darshan, soaked and freezing, stayed in the tank, with Octavia just inches from the top, Bill changed quickly into his wet suit. They prayed Octavia wouldn't try to come out.

Darshan moved back as Bill got into the cramped tank, both men careful to step around the two leather stars and the anemones on the bottom. And then, as the sea star observed the proceedings eyelessly from his usual position across from Octavia's lair, Bill bent his six-foot-five frame in half so that, although he couldn't see her tucked up under the rocks, he could feel Octavia's suckers. With his fingers, he gently urged her into his waiting net.

To Darshan's astonishment, at the touch of Bill's hand, Octavia entered Bill's net on the first try. She had not tasted Bill's skin for ten months. All that time, because she was under the ceiling of her lair, she could not see it was him handing her food on the grabber. And yet, Octavia's response to Bill's touch showed two remarkable aspects of her relationship with her keeper. She not only remembered him; she trusted him, too.

※

It's really no wonder the wild octopuses don't come out when we find them in Mooréa. Even the bolder ones—several have grabbed our pencils when prodded gently—seem to know the world is dangerous for an invertebrate without a protective shell. We've come face-to-face with a number of moray eels at our study sites, as well as sharks, and worse: We investigated one promising site and couldn't figure out why there weren't any octopuses there—until we discovered fishermen had been there before us.

So, just three days before my planned departure, most of the octopuses have done nothing but hide. When we return to one of our previous study sites to check a marked den, the octopus who lives here again shows us nothing but her suckers.

We head toward David, who is waving to us excitedly from 200 yards away. Keely and I swim slowly in the three-foot shallows, barely skimming dead corals, watching for other dens and for the puncturing spines of urchins and stonefish. I can't see the den he must be pointing to in the lumpy, algae-covered rock just one foot in front of us.

Until I see that the rock has eyes.

Arms coiled beneath its large body, the octopus sits one foot tall atop its den, its reddish, warty, nine-inch body hanging down in front of it like a big nose, its eyes camouflaged with a light starburst pattern separated by a white blaze like my border collie's. Pearly irises split by hyphen pupils swivel as it watches us. For at least a minute, the octopus remains otherwise immobile, allowing its papillae to sway limply in the surge like algae. Finally it moves. Pulling one of its arms up from beneath its body, the animal slides the arm tip inside the gill opening, as if to scratch an itch.

David and I are so mesmerized we don't even notice that Keely has swum off. Then from beneath the water, we hear her water-muffled call: "I have another one! And it's hunting!"

David stays with his octopus, and I swim over to see Keely's, just a few yards away. Again, absurdly, I cannot see it at first. At last the image my eyes must be processing materializes in my brain. This octopus is much smaller than David's, perhaps only six inches tall, even though it's standing up taller than it is wide. Mottled uniformly brown and white, its body is covered with jagged peaks of papillae, erected especially prominently above the eyes like ear tufts. If someone showed its image and relative size on a screen and asked me what animal this is, I would say it was an Eastern screech owl.

Until, with the blast of its jet, it transforms into an octopus again. Which, of course, looks like anything but an octopus. Before our eyes, Keely's octopus becomes a silken scarf, a beating heart, a gliding snail, a rock covered with algae. And then it pours itself into a hole, like water down a drain, and disappears entirely.

I pull my head out of the water and call to David. "The octopus is hunting!" I yell to him.

"Mine too!" he answers.

Keely and I now join David and begin to follow the animal as it flows over the sand on curling arms. As the octopus turns to its left, it reveals the full length of its arms—they may span more than four feet—and we glimpse a dramatic, defining moment in the animal's personal history. I can see the front three left arms all end halfway down. Like Karma, this octopus survived an encounter with a predator. The skin is healing over, but the limbs have not yet started to regrow. I feel a pang of sympathy and admiration. Surely this bold octopus remembers its brush with death, and yet it doesn't hide from us. Crawling along the bottom as we follow just feet away, it keeps us in view, seeming as curious about us as we are about it. Like us, it wants to know: *Who are you?* And for this animal, apparently, the quest to know is worth the risk. It stops, turns, and reaches out to taste my neoprene glove with its intact third right arm.

Suckers run all the way to the tip. She's a female—a swash-buckling, peg-legged pirate of an octopus, a fearless adventuress, like Kali.

We're a small, mixed-species school, following our octopus leader as she travels along the bottom. She watches us as we move with her. Abruptly, three long rows of light spots spring to her arms, while her background color changes from red to a dark brown. Then she suddenly flashes white—a behavior Jennifer has seen octopuses use to startle prey into moving. But no crab or fish appears. The color change was meant for us. Perhaps she is conducting her own personality test, the equivalent of our touching our subjects with a pencil to see what they do. But we do nothing; we keep watching. I hope our reaction doesn't disappoint her.

Next she smooths her skin, turns the color of fawn, and jets away. We frog-kick to keep up. In just a few yards, she alights on the bottom, turns chocolate, re-erects her papillae, and resumes crawling. It feels as if she is taking us, like Keith's octopus on his dive, and like Matthew's in Octopolis, on a tour of her neighborhood—a magical mystery tour, on which our guide changes shape and turns psychedelic colors. She even grows a new set of eyes. At one point of our journey together, on each side of her body, she suddenly sprouts ocelli, from the Latin for eye, two-and-a-half-inch-diameter blue rings. They signal to a predator that it has been seen, deflect attention from the real eye, and make its owner appear larger—and might have other meanings as well. At another point, she poses for David's underwater camera atop some coral rubble, inserting her arms into its holes, probing for prey. The whole time she stares ahead like a person fishing in his pocket for his keys.

Time ceased to have meaning while we swam with the octopus in the warm shallows; we could have been together for five minutes or an hour. Later we would figure the encounter lasted nearly half an hour. Eventually, David pulled his head from the water and suggested we leave so that our presence didn't interfere further with her hunting.

Our encounter with her seemed to bless our study with new luck. In the next two days, we would locate three new octopuses at this site. Ultimately, the team would find eighteen octopuses at five different sites; collect 244 shells and carapaces of octopus prey; catalog 106 food items outside active octopus dens; and identify forty-one different species of prey. Our study yielded enough data for Jennifer, David, and Tatiana to pore over happily for months.

But for me, the greatest gift of the expedition remained our swimming with the stump-armed female. David confirmed just how lucky we were. "It was definitely the best octopus encounter of my life," David said—high praise from a man who had studied octopus in the wild and in captivity for nineteen years.

Swimming with a wild octopus was a dream come true, but the octopus experience I would treasure most deeply had been back at the aquarium, with Octavia, at April's end, near the very end of her life.

⚶

Once in the barrel, Octavia had seemed very calm. She made no motions to suggest she was looking for her eggs. She didn't chew on the mesh. And Karma was delighted with her new space. She was shy at first—she wouldn't come out of her barrel until Bill led her out by one arm. But once out, she began exploring almost immediately, turning red with excitement and unfurling her body in her larger quarters like a banner in the wind.

Bill had made the right choice. Though for many months, Octavia's constant attentions to her eggs were rituals rich and full of meaning, at some point, her tending may have ceased to feel fulfilling. A wild octopus tending fertile eggs is surely rewarded, as are birds on the nest, by the signals that her eggs are alive, her embryos growing. Mother birds and their babies chirp and cheep to one another when the young are still in the egg; the mother octopus can see her babies developing inside the egg, starting with the dark eyes, and feel them moving. But Octavia had no such feedback. Perhaps the very sight of the eggs inspired her to try to protect them, the way a mother orangutan will continue to carry and even groom a dead infant, often for many days, and some dogs will refuse to leave the body of someone they love who has died. Perhaps, now that her eggs were no longer in view, Octavia at last was freed of duties she might have suspected were pointless but had felt compelled to perform. Perhaps now, at last, she could rest.

Removing Octavia from her display tank also made possible an unintentional bonus for the rest of us. When she had laid her eggs in June, we all assumed that we would never be able to touch her again. She would guard her eggs to the end, we thought, and nevermore show any interest in us. Perhaps now she would consent to touch us again—affording an opportunity for a bittersweet farewell.

When Wilson and I arrived at the aquarium the following Wednesday, we learned that Octavia hadn't moved much since the transfer. Mostly she rested in one area of the barrel, covering her swollen left eye with two of her arms. Her appetite had been steadily declining for weeks now, even though Bill had been tempting her with special delicacies; on Friday she had enjoyed a live crab, with the claws removed so it couldn't hurt her. She ate shrimp on Sunday, but nothing Monday or Tuesday.

For the first time, I dreaded seeing my old friend. All these months, I had seen her only through glass, under low light. Now, for the first time in nearly a year, I was going to see her up close again, not separated by a pane of glass. I feared what I might find. I didn't want to see her eye clouded and swollen. I didn't want to see her skin fading and thinning. I didn't want to see her weak or disoriented or upset.

Yet I longed to be with her. We hadn't touched since she had laid her eggs last June. I never knew whether, as I watched her tending her eggs from the public side of her tank, she looked out through the glass of her exhibit and recognized me as the same person who had fed her and stroked her as she looked up through the water's rippling surface so many months before.

As Christa and Brendan stood back and watched, Wilson and I unscrewed the lid from the barrel. Octavia was coiled and calm, a brownish maroon color, at the bottom. The left, swollen eye faced away from us. Her right eye, miraculously, now seemed normal, its pupil large and alert. Wilson held a squid in his right hand and waved it through the water, offering Octavia its taste and scent. Within twenty seconds she flipped upside down, and floated up, three quarters of the way to the surface, showing us her lacy white suckers. Wilson plunged his hand into the cold water to place the squid against the large suckers near her mouth. She grabbed it. I put in my hand as well, both of us offering our skin for her to taste. Would she accept us again? Would she remember?

Octavia rose a few more inches in the water, and hundreds of her suckers broke the surface. She gently grasped the back of Wilson's hand, first with just a few suckers, then with more. And then, slowly but deliberately, she extended one arm out of the water, curling it up over his hand, his wrist—and, next, her neighboring arm followed, rising and unfurling like a wine-dark wave, her suckers attaching to his hand, his wrist, his forearm.

"She knows you!" cried Christa. "She remembers you, Wilson!" And next, still embracing Wilson with two arms, she did the same to me—first with one of her arms on my right arm, then with two arms on my left. Her wet grip on my skin felt gentle and familiar, the pull of her suckers tender as a kiss.

"When I heard what happened when she recognized Bill, I almost couldn't believe it," Wilson said. "Now I do . . . there is no doubt. Clearly, she remembers."

For perhaps five minutes, Octavia stayed at the surface, holding us, tasting us, remembering us. She reached out to Christa with one arm as well—they had met once before. "What is she feeling?" I whispered.

"She is an old lady," said Wilson, tenderly, as if that observation contained the answer. Wilson grew up in a traditional culture that, unlike our own, reveres the old. In her book *The Old Way*, my friend Liz relates that the Bushmen, when approached by lions, would address them respectfully with the word *n!a*, which means "old"—"a term," Liz notes, that "they also use when speaking of the gods." The word *lady*, too, was deeply meaningful, though rarely applied to an octopus: For like a true lady, Octavia was behaving in a well-mannered and considerate way, rising to greet her friends, even though any movement must have been quite an effort.

None of us spoke as she held us, for five minutes—or was it ten? Who knew? We were on Octopus Time. Hanging upside down, Octavia offered us her white suckers, and we stroked them, and she latched on to our fingertips. Gently she blew from her siphon, but unlike the blasts of icy salt water she used to shoot, they barely rippled the surface.

Her arms were so relaxed that we could see the very tip of her beak, a black speck at the joining point of all her limbs, like the center of a flower. "She is very gentle," Wilson said softly. "She is very calm."

And then Wilson did something I had never seen him do before. He gently but deliberately put his finger to her mouth.

"I wouldn't do that, man!" warned Brendan, who had been with us the day Kali bit Anna. He'd been nearby when the arowana bit me, and had dressed my wound. Though Brendan is a tough guy who has been through a lot of physical pain himself, he hates to see other people get hurt. And Wilson is not a man to court unnecessary danger. He's not like the interns and volunteers who let the electric eel shock them just to see how it feels.

"She's not going to bite," Wilson assured Brendan. And then Wilson stroked Octavia's mouth with his index finger, petting her with an intimacy and level of trust he had never shared with any other octopus.

Finally, Octavia sank to the bottom, still regarding us with her good eye. How tired she must be, I thought, after her rich, full life—a life lived between worlds. She had known the sea's wild embrace; she had mastered the art of camouflage; she had learned the taste of our skin and the shapes of our faces; she had instinctively remembered how her ancestors wove eggs into chains. She had served as an ambassador for her kind to tens of thousands of aquarium visitors, even transforming disgust to admiration. What an odyssey she had lived.

I leaned over the barrel and stared at her in awe and gratitude.

My eyes brimmed, and a tear dripped into the water. Human tears of intense emotion are chemically distinct from tears produced by eye irritants; tears of both joy and sorrow contain prolactin, a hormone that peaks in men and women during sex, dreams, and seizures, and is associated in women with the synthesis of breast milk. I wondered if Octavia could taste my feelings. She may have recognized the taste: Fish have prolactin. Octopuses do too.

As she rested, pale areas spread over Octavia's brownish skin in a weblike pattern. "She's beautiful," said Brendan reverently. He had never seen her up close before, only through the glass of her exhibit. But even now, even so near the end of her life, Octavia *was* beautiful, and except for her bad eye, she looked healthy, if thin, with no white patches of dead skin. "She's a beautiful old lady," I said.

We rested, people and octopus regarding one another for several more minutes. And then, to my surprise, she floated up again to us. As she rose, we scanned the bottom of the barrel and saw that she had dropped the squid Wilson had fed her. We wished she had eaten, but we learned something new: Hunger was not the reason she had surfaced earlier, and it wasn't what brought her now.

The reason she surfaced was abundantly clear. She had not interacted with us, or tasted our skin, or seen us above her tank for ten full months. She was sick and weak. In less than four weeks, on a Saturday morning in May, Bill would find her, pale, thin, and still, dead at the bottom of her barrel. Yet, despite everything, we knew in that moment that Octavia had not only remembered us and recognized us; she had wanted to touch us again.

⚜

When it was first filled with harbor water, the finished Giant Ocean Tank glowed a magical, promising, living green. As it cleared, like dawn, its waters sparkled with the colors and shapes of all the new

coral sculptures and hundreds of new fishes. Staff had wisely returned the smallest fish to the tank first, so they could own the crevices, establishing territories where they'd feel safe and confident,
before introducing the larger, predatory fishes—who never bothered
them. Myrtle again ruled her old domain. The penguins returned to
their tray, took up the exact same spots each had defended eleven
months before, and again filled the aquarium's first floor with their
welcome, raucous braying.

The public opening for the transformed GOT on July 1 was
glorious—and so was Marion and Dave's wedding. Scott's secular
homily focused on Marion's work with anacondas, prompting one
guest to comment, "Well, that was the best *snake* wedding I ever
attended." Wilson and his family managed to spring his wife from
her assisted living facility to attend, in her wheelchair, a party for
her granddaughter, Sophie. She recognized everyone and seemed to
enjoy the gathering. At the end of the summer, Christa was chosen
out of fifty people who were interviewed for the one open permanent position in the aquarium's education department. Four hundred and thirty thousand people came to visit the New England
Aquarium that July and August, the highest number of visitors in
the institution's forty-four-year history.

❧

On a Wednesday in September, I arrive at the octopus tank to find
Karma entertaining a crowd, crawling vigorously across the front
of her exhibit on her large, white suckers, watching the people out
of one slit-pupil eye. "Ooooh! Oct-oh-PUS!" cries a little girl with
three blond pigtails held in place with pink ribbons. "Wow! Way
cool!" says a teenage boy in a leather jacket. "Come see this, class!"
a school trip chaperone summons the students in her charge. "The
octopus is out!"

I rush upstairs to join Wilson, and we open the top of her tank. Karma, bright red, flows over to greet us and flips upside down. We can see the flash of cameras below as we hand her one capelin after another, six in all, which she accepts eagerly. We see the fish crowded together at the center of her arms. But even as she's eating, she's eager to play with us. Her arms coil out of the water to grab us, sucking so vehemently she leaves love bites on our forearms and the backs of our hands.

Did she actually eat all six capelin? We go back downstairs to her exhibit to see if she dropped any.

"Was that you with the octopus?" a young boy asks, as if we had just been seen having dinner with the president. We nod proudly.

"Does it *know* you?" a middle-aged man with a moustache asks, incredulous.

Of course she does, we answer. Perhaps as well as or better than we know her.

But I still have so many questions. What goes on in Karma's head—or the larger bundle of neurons in her arms—when she sees us? Do her three hearts beat faster when she catches sight of Bill, or Wilson, or Christa, or Anna, or me? Would she feel sad if we disappeared? What does sadness feel like for an octopus—or for anyone else, for that matter? What does Karma feel like when she pours her huge body into a tiny crevice of her lair? What does capelin taste like on her skin?

I can't know this, of course; and I can't know exactly what I mean to her. But I know what she—and Octavia and Kali—have meant to me. They have changed my life forever. I loved them, and will love them always, for they have given me a great gift: a deeper understanding of what it means to think, to feel, and to know.

Postscript

In April of 2016, the New England Aquarium celebrates the opening of a new and much larger octopus exhibit, replacing the 560-gallon tank where Athena, Octavia, Karma, and their predecessors lived. All the inhabitants of three separate exhibits, including sea stars, wolf eels, giant green anemones, and several rock fish, will now live together in the new, 6,000-gallon tank. The old tank was roughly the size of a VW Beetle while the new tank is more the size of a truck. Within the huge new tank, two separate "apartments"—each more than twice the size of the old one—will house an octopus, allowing them to see each other while keeping them safely apart.

Now, when a second octopus-in-waiting is in residence, he or she will no longer live behind the scenes in a barrel but in a roomy octopus apartment. Visitors as well as the "star" octopus in the front tank will be able to see the younger octopus.

Will the octopuses like the new arrangement? "We're hoping so," Bill Murphy says. He has been dreaming about this tank redesign since 2007. "We'll not only have a larger space but also new animals for her," he told me a few months before the new tank was to open. "Or him." The star octopus will still share quarters with the large male sunflower sea star and other stars, but now they'll be joined by other creatures as well. "She'll be able to reach out and touch the sea pens," Bill said. "We'll see if she likes that."

Both octopus apartments feature new rockwork with even more nooks and crannies for eager arms to stay busy exploring. And though human interaction with the octopuses will continue, the octopuses will now also be able to spend time with even more of the species they would live with in the wild—and, as Bill notes proudly, "give visitors a great peek into the waters of the Pacific Northwest." In some cases, the other creatures will be separated from the octopus by plexiglass panels, so the octopuses won't eat them.

Acknowledgments

I am deeply indebted to everyone, vertebrate and invertebrate, mentioned in these pages, and to many who are not.

Without exception, each volunteer and staff member I have met at the New England Aquarium has been extraordinarily knowledgeable and helpful, welcoming me to explore the institution's every aspect. Anita Metzler showed me her research lobsters (each with a distinctive personality). Jamie Mathison introduced me to the aquarium's seals and sea lions, and had Amelia the harbor seal kiss my lips and Cordova the northern fur seal kiss my nose. John Reardon, senior watch engineer, gave me a tour of the basement, the humming heart of the aquarium, with its giant storage tanks and pumps and filters. These and many other scenes never made it into this book, but I will remember them, and the kindness of those who made them possible, always.

Likewise I will never forget my visit to the octopus lab at Middlebury College in Vermont and its eleven California two-spot octopuses, especially the outgoing and vivacious female, Octopus 1. Many thanks to director of neuroscience Tom Root, director of animal research Vicki Major, assistant animal caretaker Carolyn Clarkson, and the other caretakers and researchers who welcomed me and my friends from the New England Aquarium on our daylong tour there.

Several expeditions further afield, one above water and two below, were critical to this book. My two research trips to the Seattle Aquarium, to attend the Octopus Symposium and the Octopus Blind Date, were rendered especially informative and meaningful thanks to the kindness of the late Roland Anderson. I am grateful to the staff of United Divers and to Barb Sylvestre at Aquatic Specialties for opening to me the world of diving. I thank scuba instructor extraordinaire Doris Morrissette for including me on the trip to Cozumel, Mexico, where I saw my first wild octopuses. I thank, too, the other people in our tight-knit group: Rob Sylvestre, Walter Hooker, Mary Ann Johnston, Mike Beresford, and Janice and Ray Nadeau, as well as the helpful and knowledgeable staff of Scuba Club Cozumel. For her pioneering research, generous spirit, and enduring friendship, I thank Jennifer Mather, who organized and led the octopus research expedition to Mooréa, French Polynesia, with fellow researchers David Scheel, Tatiana Leite, and Keely Langford, friends whom I will always treasure.

In addition, I owe special thanks to:

Orion editor (and wonderful author) Andrew Blechman for his encouragement for my original article on octopuses in his fine magazine.

Evolutionary biologist Gary Galbreath, my chief scientific adviser and hero, gifted teacher, respected field researcher, and friend to animals everywhere.

Jody Simpson, for listening to this book evolve over the course of three years and many hundreds of long walks in the woods with our dogs, Sally, Pearl, and May.

Polymath Mike Strzelec, indestructible adventurer who ably served as a research assistant while I was writing this book, sending me octopodan articles on topics ranging from the function of cephalopod nidamental glands to the less-welcome Japanese concept of "tentacle sex."

Hancock, New Hampshire, librarian Amy Markus, who did double duty as my personal assistant while I was away researching this book.

Author and translator Jerry Ryan, for his insights into the New England Aquarium's rich history.

Tianne Strombeck, for her sensitive and revealing portraits of Octavia, Kali, and Karma; Johanna Blasi, for her photos of the Giant Ocean Tank; and Keith Ellenbogen and David Scheel, for their splendid company and photos from our trip to Mooréa.

Designer Paul Dippolito for the elegant look of this book. Liz Thomas, for her unending support and flawless advice as I researched and wrote this book—and all the books before and after this one.

My husband, Howard Mansfield, for his exceptional editorial scrutiny, patience, and kindness, and for putting up with all those octopus hickeys.

Marion Britt, Christa Carceo, Selinda Chiquoine, Marc Dohan, Scott Dowd, Joel Glick, Jennifer Mather, Marion and Sam Magill-Dohan, Robert Matz, Wilson Menashi, Bill Murphy, Andrew Murphy, Judith and Robert Oksner, Jerry Price, Liz Thomas, Jody Simpson, Gretchen Vogel, and Polly Watson for carefully reading, correcting, and commenting on the manuscript.

My beloved literary agent, Sarah Jane Freymann, my terrific editor, Leslie Meredith, and her wonderful associate editor, Donna Loffredo, who all believed in this book from the start.

Selected Bibliography

The following is a list of some of the books, articles, videos, and websites I found especially helpful when thinking about and researching this book:

BOOKS

Bailey, Elisabeth Tova. *The Sound of a Wild Snail Eating.* Chapel Hill, NC: Algonquin Books of Chapel Hill, 2010.

Blackmore, Susan. *Consciousness: A Very Short Introduction.* Oxford, UK: Oxford University Press, 2005.

Cosgrove, James A., and Neil McDaniel. *Super Suckers: The Giant Pacific Octopus and Other Cephalopods of the Pacific Coast.* Madeira Park, BC: Harbour Press, 2009.

Courage, Katherine Harmon. *Octopus! The Most Mysterious Creature in the Sea.* New York: Penguin, 2013.

Cousteau, Jacques, and Philippe Diolé. *Octopus and Squid: The Soft Intelligence.* New York: Doubleday, 1973.

Damasio, Antonio. *The Feeling of What Happens: Body and Emotion in the Making of Consciousness.* New York: Harcourt Brace and Co., 1999.

Dennett, Daniel C. *Kinds of Minds: Toward an Understanding of Consciousness.* New York: Basic Books, 1996.

Dunlop, Colin, and Nancy King. *Cephalopods: Octopuses and Cuttlefishes for the Home Aquarium.* Neptune City, NJ: TFH Publications, 2009.

Ellis, Richard. *The Search for the Giant Squid: The Biology and Mythology of the World's Most Elusive Sea Creature.* New York: Penguin, 1999.

Fortey, Richard. *Horseshoe Crabs and Velvet Worms: The Story of the Animals and Plants That Time Has Left Behind.* New York: Knopf, 2012.

Foulkes, David. *Children's Dreaming and the Development of Consciousness.* Cambridge, MA: Harvard University Press, 1999.

Gibson, James William. *A Reenchanted World: The Quest for a New Kinship with Nature.* New York: Holt, 2009.

Gomez, Luiz. *A Pictorial Guide to Common Fish in the Mexican Caribbean.*

Cancún, Mexico: Editora Fotografica Marina Kukulcan S.A. de C.V., 2012.

Grant, John, and Ray Jones. *Window to the Sea*. Guilford, CT: Globe Pequot Press, 2006.

Gregg, Justin. *Are Dolphins Really Smart? The Mammal Behind the Myth*. Oxford, UK: Oxford University Press, 2013.

Hall, James A. *Jungian Dream Interpretation*. Toronto: Inner City Books, 1983.

Humann, Paul, and Ned Deloach. *Reef Creature Identification: Florida Caribbean Bahamas*. Jacksonville, FL: New World Publications, 2002.

_____. *Reef Coral Identification: Florida Caribbean Bahamas*. Jacksonville, FL: New World Publications, 2011.

Jaynes, Julian. *The Origin of Consciousness in the Breakdown of the Bicameral Mind*. Boston: Houghton Mifflin, 1976.

Keenan, Julian Paul. *The Face in the Mirror: The Search for Origins of Consciousness*. New York: Harper Collins Ecco, 2003.

Lane, Frank. *Kingdom of the Octopus*. New York: Pyramid Publications, 1962.

Lewbel, George S., and Larry R. Martin. *Diving and Snorkeling Cozumel*. St. Footscray, Victoria, Australia: Lonely Planet Publications, 2006.

Linden, Eugene. *The Octopus and the Orangutan*. New York: Dutton, 2002.

Mather, Jennifer, Roland C. Anderson, and James B. Wood. *Octopus: The Ocean's Intelligent Invertebrate*. Portland, OR: Timber Press, 2010.

Mather, J. A. "Cephalopod Displays: From Concealment to Communication." In *Evolution of Communication Systems*, eds. D. Kimbrough Oller and Ulrike Griebel, 193–213. Cambridge, MA: MIT Press, 2004.

Morell, Virginia. *Animal Wise: The Thoughts and Emotions of Our Fellow Creatures*. New York: Crown, 2013.

Moynihan, Martin. *Communication and Noncommunication by Cephalopods*. Bloomington, IN: Indiana University Press, 1985.

Paust, Brian C. *Fishing for Octopus: A Guide for Commercial Fishermen*. Fairbanks, AK: Sea Grant/University of Alaska, 2000.

Prager, Ellen. *Sex, Drugs, and Sea Slime: The Oceans' Oddest Creatures and Why They Matter*. Chicago: University of Chicago Press, 2012.

Ryan, Jerry. *A History of the New England Aquarium 1957–2004*. Boston: produced for limited distribution by the author, 2011.

_____. *The Forgotten Aquariums of Boston*, 2nd rev. ed. Pascoag, RI: Finley Aquatic Books, 2002.

Segaloff, Nat, and Paul Erickson. *A Reef Comes to Life: Creating an Undersea Exhibit*. Boston: Franklin Watts, 1991.

Shubin, Neil. *Your Inner Fish: A Journey into the 3.5-Billion-Year History of the Human Body*. New York: Vintage, 2009.

Siers, James. *Moorea*. Wellington, New Zealand: Millwood Press, 1974.

Williams, Wendy. *Kraken: The Curious, Exciting, and Slightly Disturbing Science of Squid*. New York: Abrams Image, 2011.

SCIENTIFIC ARTICLES

Anderson, Roland D., Jennifer Mather, Mathieu Q. Monette, and Stephanie R. M. Zimsen. "Octopuses (*Enteroctopus Doflenini*) Recognize Individual Humans." 2010. *Journal of Applied Animal Welfare Science* 13: 261–72.

Boal, Jean Geary, Andrew W. Dunham, Kevin T. Williams, and Roger T. Hanlon. "Experimental Evidence for Spatial Learning in Octopuses (*Octopus Biomaculoides*)." 2000. *Journal of Comparative Psychology* 114: 246–52.

Brembs, B. "Towards a Scientific Concept of Free Will as a Biological Trait: Spontaneous Actions and Decision-Making in Invertebrates." 2011. *Proceedings of the Royal Society of Biological Sciences* 278 (170): 930–39.

Byrne, Ruth, Michael J. Kuba, Daniela V. Meisel, Ulrike Griebel, and Jennifer Mather. "Does *Octopus Vulgaris* Have Preferred Arms?" 2006. *Journal of Comparative Psychology* 120 (3): 198–204.

Godfrey-Smith, Peter, and Matthew Lawrence. "Long-Term High-Density Occupation of a Site by *Octopus Tetricus* and Possible Site Modification Due to Foraging Behavior." 2012. *Marine Freshwater Behavior and Physiology* 45 (4): 261–68.

Hochner, Binyamin, Tal Shormrat, and Graziano Fiorito. "The Octopus: A Model for a Comparative Analysis of the Evolution of Learning and Memory Mechanisms." 2006. *The Biological Bulletin* 210 (3): 308–17.

Leite, T. S., M. Haimovici, W. Molina, and K. Warnke. "Morphological and Genetic Description of *Octopus Insularis*, a New Cryptic Species of the *Octopus Vulgaris* Complex from the Tropical Southwestern Atlantic." 2008. *Journal of Molluscan Studies* 74 (1): 63–74.

Lucerno, M., H. Farrington, and W. Gilly. "Quantification of L-Dopa and Dopamine in Squid Ink: Implications for Chemoreception." 1994. *The Biological Bulletin* 187 (1): 55–63.

Mather, J. A., Tatiana Leite, and Allan T. Battista. "Individual Prey Choices of Octopus: Are They Generalists or Specialists?" 2012. *Current Zoology* 58 (4): 597–603.

Mather, J. A. "Cephalopod Consciousness: Behavioral Evidence." 2008. *Consciousness and Cognition* 17 (1): 37–48.

Mather, J. A., and Roland C. Anderson. "Ethics and Invertebrates: a Cephalopod Perspective." 2007. *Diseases of Aquatic Organisms* 75: 119-129.

Mather, J. A., and R. C. Anderson. "Exploration, Play and Habituation in *Octopus Dofleini.*" 1999. *Journal of Comparative Psychology* 113: 333–38.

Mather, Jennifer A. "Cognition in Cephalopods." 1995. *Advances in the Study of Behavior* 24: 316–53.

Mather, J. A. "'Home' Choice and Modification by Juvenile *Octopus Vulgaris*: Specialized Intelligence and Tool Use?" 1994. *Journal of Zoology* (London) 233: 359–68.

Mathger, Lydia M., Steven B. Roberts, and Roger T. Hanlon. "Evidence for Distributed Light Sensing in the Skin of Cuttlefish, *Sepia Officinalis.*" 2011. *Biology Letters* 6: 600–03.

Nair, J. Rajasekharan, Devika Pillai, Sophia Joseph, P. Gomathi, Priya V. Senan, and P. M. Sherief. "Cephalopod Research and Bioactive Substances." 2011. *Indian Journal of Geo-Marine Sciences* 40 (1): 13–27.

Toussaint, R. K., David Scheel, G. K. Sage, and S. L. Talbot. "Nuclear and Mitochondrial Markers Reveal Evidence for Genetically Segregated Cryptic Speciation in Giant Pacific Octopuses from Prince William Sound, Alaska." 2012. *Conservation Genetics* 13 (6): 1483–97.

VIDEOS ON THE WEB

Web addresses frequently change; I used these when researching the book.

New England Aquarium's Bill Murphy interacts with his late friend George, a giant Pacific octopus:

http://www.youtube.com/watch?v=_60WQZkgiaU

An octopus materializes from a piece of algae—and then swims away:

http://www.youtube.com/watch?v=ckP8msIgMYE

The first of Roger Hanlon's excellent series of lectures on camouflage and

signaling in cephalopods. Follow the links to watch the others:
http://www.youtube.com/watch?v=oDvvVOlyaLI

A diver was shocked when an octopus seized his new video camera from his hands—and then made off with the prize, the camera rolling all the while:
http://www.youtube.com/watch?v=x5DyBkYKqnM

Releasing the Dude, a giant Pacific octopus who briefly lived at the Shaw Ocean Discovery Centre, into the wild:
http://www.youtube.com/watch?v=V57Dfn_F69c

When sharks started showing up dead at the Seattle Aquarium's large tank, the octopus was found to blame:
http://www.youtube.com/watch?v=urkC8pLMbh4

Training lumpfish at the New England Aquarium:
http://www.youtube.com/watch?v=7j9SovBHpUw

A veined octopus tiptoes the sea bottom while carrying a coconut shell half as portable armor:
http://www.news.nationalgeographic.com/news/2009/12/091214-octopus-carries-coconuts-coconut-carrying.html

In a cove in West Seattle, a giant Pacific octopus tenderly cares for her 50,000 eggs as they hatch during the last week of her life:
http://www.huffingtonpost.com/2011/06/23/giant-pacific-octopus-bab_n_883384.html

AUDIO

The team of Living on Earth visits Octavia at the New England Aquarium:
http://www.youtube.com/watch?v=7j9SovBHpUw

OTHER ONLINE RESOURCES

New England Aquarium's Web page, with visitor information, videos, news, and special programs: www.neaq.org

Seattle Aquarium's Web page, which will include announcements of Octopus Week and the biannual Octopus Symposium: www.seattleaquarium.org

The Octopus News Magazine Online: News about octopus, nautilus, squid, cuttlefish, and cephalopod fossils: www.TONMO.com

Philosopher-diver Peter Godfrey-Smith's fascinating blog about cephalopod evolution, bodies, minds, and the sea, including a special section on Octopolis: www.giantcuttlefish.com

MD/PhD student Mike Lisieski's excellent, in-depth articles and video on cephalopods, with special emphasis on intelligence and camouflage: cephalove.southernfriedscience.com

Author and *Scientific American* contributing editor Katherine Harmon Courage's clever and breezy blog about octopuses: http://blogs.scientificamerican.com/octopus-chronicles/

Index

About the Author

Sy Montgomery is the author of twenty acclaimed books of nonfiction for adults and children. Her memoir for adults, *The Good Good Pig*, was a national best seller. Her *Kakapo Rescue* won the Robert F. Sibert Informational Book Medal, the highest award given specifically for a book of children's nonfiction. Montgomery's *Spell of the Tiger* was the inspiration for a National Geographic TV documentary of the same name about her work. Her Amazon adventure, *Journey of the Pink Dolphins*, was called "rhapsodic" (*Publishers Weekly*), "mesmerizing" (*Booklist*), and "searching and personal" (*The New Yorker*), and was a finalist for the London Times Travel Book Award. She is the recipient of numerous other honors, including lifetime achievement awards from the Humane Society of the United States and the New England Booksellers Association, and three honorary degrees.

Sy lives in New Hampshire with her border collie, Sally, her free-range flock of laying hens, and her husband, the writer Howard Mansfield.

Other Books by Sy Montgomery